·梁 灏 编著·

Vue.js 实战

清华大学出版社
北京

内 容 简 介

本书以 Vue.js 2 为基础，以项目实战的方式来引导读者渐进式学习 Vue.js。本书分为基础篇、进阶篇和实战篇三部分。基础篇主要是对 Vue.js 核心功能的介绍；进阶篇主要讲解前端工程化 Vue.js 的组件化、插件的使用；实战篇着重开发了两个完整的示例，所涉及的内容涵盖 Vue.js 绝大部分 API。通过阅读本书，读者能够掌握 Vue.js 框架主要 API 的使用方法、自定义指令、组件开发、单文件组件、Render 函数、使用 webpack 开发可复用的单页面富应用等。

本书示例丰富、侧重实战，适用于刚接触或即将接触 Vue.js 的开发者，也适用于对 Vue.js 有过开发经验，但需要进一步提升的开发者。

本书封面贴有清华大学出版社防伪标签，无标签者不得销售。
版权所有，侵权必究。举报：010-62782989，beiqinquan@tup.tsinghua.edu.cn。

图书在版编目（CIP）数据

Vue.js 实战/梁灏编著. —北京：清华大学出版社，2017（2024.2重印）
ISBN 978-7-302-48492-9

Ⅰ. ①V… Ⅱ. ①梁… Ⅲ. ①网页制作工具－程序设计 Ⅳ. ①TP392.092.2

中国版本图书馆 CIP 数据核字（2017）第 230264 号

责任编辑：王金柱
封面设计：王　翔
责任校对：闫秀华
责任印制：宋　林

出版发行：清华大学出版社
网　　址：https://www.tup.com.cn，https://www.wqxuetang.com
地　　址：北京清华大学学研大厦 A 座　　　　　邮　　编：100084
社 总 机：010-83470000　　　　　　　　　　　邮　　购：010-62786544
投稿与读者服务：010-62776969，c-service@tup.tsinghua.edu.cn
质 量 反 馈：010-62772015，zhiliang@tup.tsinghua.edu.cn

印 装 者：三河市铭诚印务有限公司
经　　销：全国新华书店
开　　本：190mm×260mm　　印　张：21.5　　字　数：550 千字
版　　次：2017 年 10 月第 1 版　　　　　　　　印　次：2024 年 2 月第 17 次印刷
定　　价：79.00 元

产品编号：074441-01

推 荐 序

在撰写 Vue 文档的过程中，出于篇幅和精力的限制，主要着力于对 Vue 本身 API 的解释。对于缺乏实战经验的读者来说，虽然可能明白了 API 的用法，但对于如何将它使用在实际项目中仍然会感到困惑。而这本书的优点，正是在于对重要的知识点结合了一些实战范例来帮助读者更好地理解 API 设计的初衷和使用场景，并且在 GitHub 有对应的源码可以下载研究。

本书的作者梁灏是优秀的开源 Vue 组件库 iView 的作者，也为 Vue 社区的活跃做出了很多贡献。同时，对开源的投入也使得他对 Vue 的设计和底层实现有相当深入的理解。

如果你喜欢通过实例来学习，那么这本书会是你上手 Vue 的一个好选择。

尤雨溪
2017 年 6 月

引　言

两年前，我开始接触 Vue.js 框架，当时就被它的轻量、组件化和友好的 API 所吸引。之后我将 Vue.js 和 webpack 技术栈引入我的公司（TalkingData）可视化团队，并经过一年多的实践，现已成为整个公司的前端开发规范。

与此同时，我开源了 iView 项目，它是基于 Vue.js 的一套高质量 UI 组件库，从设计规范、工程构建到国际化都提供了完整的解决方案，并支持 SSR。在许多志愿者的帮助下，将文档全部翻译为英文，在 Vue 开发者社区颇受欢迎。

如今，前端框架可谓百家争鸣，但每一个框架的产生都是为了解决具体问题的。Vue.js 以渐进式切入，对不同阶段的开发者提供了不同的开发模式，由浅入深。Vue.js 提供了友好的 API 和强大的功能，包括双向数据绑定、路由、状态管理、动画、组件化、SSR，无论是简单的页面还是复杂的系统，从可复用性、便捷性和维护性上都精益求精。

有幸完成此书，希望能给 Vue.js 社区带来一点帮助。

梁灏（Aresn）

2017 年 7 月 10 日

关于本书

本书分为"基础篇""进阶篇"和"实战篇"三部分,其中基础篇涵盖了 Vue.js 2 的所有基础内容,包括:

- 双向绑定数据;
- 计算属性;
- 内置指令与自定义指令;
- 组件。

基础篇内容相对容易,适合刚入门 Vue.js 的开发者。

进阶篇更深入 Vue.js 的工程化,内容包括:

- Render 函数;
- webpack 的使用;
- Vue.js 插件。

实战篇首先剖析了 iView 的两个经典组件的设计和实现思路,然后充分利用 Vue.js 的内容完成了两个完整的实战项目。

本书读者对象

本书基础内容和实战项目共存,适用于刚接触 Vue.js 的前端或后端开发者。当然,有一定 Vue.js 开发经验的读者也能从中收获不少实战中的经验。

本书要求读者有一定的 JavaScript 编程能力,并假定读者已经掌握了 HTML 和 CSS,否则阅读本书会有一定的难度。

其他说明

本书基础篇示例只需要有编辑器和浏览器两个环境即可,浏览器以 Chrome 最佳,如果是 IE,至少在 IE 9 版本以上。进阶篇和实战篇的示例需要读者了解并安装 Node.js 和 NPM,本书将不会介绍它们的安装方法。

本书相关的示例代码都放在 GitHub,网址如下:

https://github.com/icarusion/vue-book

读者可到该项目的 issues 讨论区提问讨论。

目　　录

第 1 篇　基础篇

第 1 章　初识 Vue.js 3

1.1　Vue.js 是什么 3
　　1.1.1　MVVM 模式 3
　　1.1.2　Vue.js 有什么不同 4
1.2　如何使用 Vue.js 5
　　1.2.1　传统的前端开发模式 5
　　1.2.2　Vue.js 的开发模式 5

第 2 章　数据绑定和第一个 Vue 应用 8

2.1　Vue 实例与数据绑定 9
　　2.1.1　实例与数据 9
　　2.1.2　生命周期 10
　　2.1.3　插值与表达式 11
　　2.1.4　过滤器 13
2.2　指令与事件 15
2.3　语法糖 18

第 3 章　计算属性 19

3.1　什么是计算属性 19
3.2　计算属性用法 20
3.3　计算属性缓存 23

第 4 章　v-bind 及 class 与 style 绑定 25

4.1　了解 v-bind 指令 25
4.2　绑定 class 的几种方式 26
　　4.2.1　对象语法 26
　　4.2.2　数组语法 27
　　4.2.3　在组件上使用 29
4.3　绑定内联样式 30

第 5 章 内置指令 ... 32

5.1 基本指令 ... 32
5.1.1 v-cloak ... 32
5.1.2 v-once ... 33

5.2 条件渲染指令 ... 33
5.2.1 v-if、v-else-if、v-else ... 33
5.2.2 v-show ... 36
5.2.3 v-if 与 v-show 的选择 ... 36

5.3 列表渲染指令 v-for ... 37
5.3.1 基本用法 ... 37
5.3.2 数组更新 ... 41
5.3.3 过滤与排序 ... 43

5.4 方法与事件 ... 44
5.4.1 基本用法 ... 44
5.4.2 修饰符 ... 46

5.5 实战：利用计算属性、指令等知识开发购物车 ... 47

第 6 章 表单与 v-model ... 55

6.1 基本用法 ... 55
6.2 绑定值 ... 61
6.3 修饰符 ... 63

第 7 章 组件详解 ... 65

7.1 组件与复用 ... 65
7.1.1 为什么使用组件 ... 65
7.1.2 组件用法 ... 66

7.2 使用 props 传递数据 ... 70
7.2.1 基本用法 ... 70
7.2.2 单向数据流 ... 72
7.2.3 数据验证 ... 74

7.3 组件通信 ... 75
7.3.1 自定义事件 ... 75
7.3.2 使用 v-model ... 77
7.3.3 非父子组件通信 ... 79

7.4 使用 slot 分发内容 ... 83
7.4.1 什么是 slot ... 83
7.4.2 作用域 ... 84

		7.4.3 slot 用法	85
		7.4.4 作用域插槽	87
		7.4.5 访问 slot	89
	7.5	组件高级用法	90
		7.5.1 递归组件	90
		7.5.2 内联模板	92
		7.5.3 动态组件	93
		7.5.4 异步组件	94
	7.6	其他	95
		7.6.1 $nextTick	95
		7.6.2 X-Templates	96
		7.6.3 手动挂载实例	97
	7.7	实战：两个常用组件的开发	98
		7.7.1 开发一个数字输入框组件	98
		7.7.2 开发一个标签页组件	106

第 8 章 自定义指令 ... 118

8.1	基本用法	118
8.2	实战	121
	8.2.1 开发一个可从外部关闭的下拉菜单	121
	8.2.2 开发一个实时时间转换指令 v-time	126

第 2 篇　进阶篇

第 9 章 Render 函数 ... 133

9.1	什么是 Virtual Dom	133
9.2	什么是 Render 函数	136
9.3	createElement 用法	140
	9.3.1 基本参数	140
	9.3.2 约束	143
	9.3.3 使用 JavaScript 代替模板功能	147
9.4	函数化组件	153
9.5	JSX	157
9.6	实战：使用 Render 函数开发可排序的表格组件	159
9.7	实战：留言列表	172
9.8	总结	183

第 10 章 使用 webpack 184

10.1 前端工程化与 webpack 184
10.2 webpack 基础配置 187
10.2.1 安装 webpack 与 webpack-dev-server 187
10.2.2 就是一个 js 文件而已 188
10.2.3 逐步完善配置文件 191
10.3 单文件组件与 vue-loader 194
10.4 用于生产环境 201

第 11 章 插件 206

11.1 前端路由与 vue-router 207
11.1.1 什么是前端路由 207
11.1.2 vue-router 基本用法 208
11.1.3 跳转 212
11.1.4 高级用法 213
11.2 状态管理与 Vuex 216
11.2.1 状态管理与使用场景 216
11.2.2 Vuex 基本用法 217
11.2.3 高级用法 221
11.3 实战：中央事件总线插件 vue-bus 227

第 3 篇　实战篇

第 12 章 iView 经典组件剖析 235

12.1 级联选择组件 Cascader 236
12.2 折叠面板组件 Collapse 249
12.3 iView 内置工具函数 257

第 13 章 实战：知乎日报项目开发 261

13.1 分析与准备 261
13.2 推荐列表与分类 265
13.2.1 搭建基本结构 265
13.2.2 主题日报 267
13.2.3 每日推荐 271
13.2.4 自动加载更多推荐列表 276
13.3 文章详情页 278

		13.3.1 加载内容	278
		13.3.2 加载评论	281
	13.4	总结	286

第 14 章 实战：电商网站项目开发 288

14.1	项目工程搭建	288
14.2	商品列表页	290
	14.2.1 需求分析与模块拆分	290
	14.2.2 商品简介组件	291
	14.2.3 列表按照价格、销量排序	297
	14.2.4 列表按照品牌、颜色筛选	306
14.3	商品详情页	309
14.4	购 物 车	313
	14.4.1 准备数据	314
	14.4.2 显示和操作数据	316
	14.4.3 使用优惠码	320
14.5	总结	324

第 15 章 相关开源项目介绍 325

15.1	服务端渲染与 Nuxt.js	325
	15.1.1 是否需要服务端渲染	325
	15.1.2 Nuxt.js	326
15.2	HTTP 库 axios	327
15.3	多语言插件 vue-i18n	329

第 1 篇 基础篇

基础篇将循序渐进地介绍 Vue.js 的核心功能，包括数据的双向绑定、计算属性、基本指令、自定义指令及组件等。通过对基础篇的学习，可以快速构建出 Vue.js 应用并直接用于生产环境。

第 1 章

初识 Vue.js

本章主要介绍与 Vue.js 有关的一些概念与技术，并帮助你了解它们背后相关的工作原理。通过对本章的学习，即使从未接触过 Vue.js，你也可以运用这些知识点快速构建出一个 Vue.js 应用。

1.1 Vue.js 是什么

Vue.js 的官方文档中是这样介绍它的。

简单小巧的核心，渐进式技术栈，足以应付任何规模的应用。

简单小巧是指 Vue.js 压缩后大小仅有 17KB。所谓渐进式（Progressive），就是你可以一步一步、有阶段性地来使用 Vue.js，不必一开始就使用所有的东西。随着本书的不断介绍，你会深刻感受到这一点，这也正是开发者热爱 Vue.js 的主要原因之一。

使用 Vue.js 可以让 Web 开发变得简单，同时也颠覆了传统前端开发模式。它提供了现代 Web 开发中常见的高级功能，比如：

- 解耦视图与数据。
- 可复用的组件。
- 前端路由。
- 状态管理。
- 虚拟 DOM（Virtual DOM）。

1.1.1 MVVM 模式

与知名前端框架 Angular、Ember 等一样，Vue.js 在设计上也使用 MVVM（Model-View-View Model）模式。

MVVM 模式是由经典的软件架构 MVC 衍生来的。当 View（视图层）变化时，会自动更新到 ViewModel（视图模型），反之亦然。View 和 ViewModel 之间通过双向绑定（data-binding）建立联系，如图 1-1 所示。

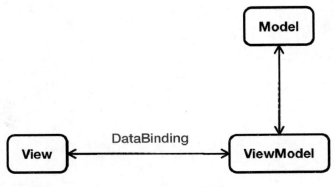

图 1-1　MVVM 关系

1.1.2　Vue.js 有什么不同

如果你使用过 jQuery，那你一定对操作 DOM、绑定事件等这些原生 JavaScript 能力非常熟悉，比如我们在指定 DOM 中插入一个元素，并给它绑定一个点击事件：

```
if (showBtn) {
    var btn = $('<button>Click me</button>');
    btn.on('click', function () {
        console.log('Clicked!');
    });
    $('#app').append(btn);
}
```

这段代码不难理解，操作的内容也不复杂，不过这样让我们的视图代码和业务逻辑紧耦合在一起，随着功能不断增加，直接操作 DOM 会使得代码越来越难以维护。

而 Vue.js 通过 MVVM 的模式拆分为视图与数据两部分，并将其分离。因此，你只需要关心你的数据即可，DOM 的事情 Vue 会帮你自动搞定，比如上面的示例用 Vue.js 可以改写为：

```
<body>
    <div id="app">
        <button v-if="showBtn" v-on:click="handleClick">Click me</button>
    </div>
</body>
<script>
    new Vue({
        el: '#app',
        data: {
            showBtn: true
        },
```

```
        methods: {
            handleClick: function () {
                console.log('Clicked!');
            }
        }
    })
</script>
```

提示　暂时还不需要理解上述代码，这里只是快速展示 Vue.js 的写法，在后面的章节会详细介绍每个参数的用法。

1.2　如何使用 Vue.js

每一个框架的产生都是为了解决某个具体的问题。在正式开始学习 Vue.js 前，我们先对传统前端开发模式和 Vue.js 的开发模式做一个对比，以此了解 Vue.js 产生的背景和核心思想。

1.2.1　传统的前端开发模式

前端技术在近几年发展迅速，如今的前端开发已不再是 10 年前写个 HTML 和 CSS 那样简单了，新的概念层出不穷，比如 ECMAScript 6、Node.js、NPM、前端工程化等。这些新东西在不断优化我们的开发模式，改变我们的编程思想。

随着这些技术的普及，一套可称为"万金油"的技术栈被许多商业项目用于生产环境：

jQuery + RequireJS（SeaJS）+ artTemplate（doT）+Gulp（Grunt）

这套技术栈以 jQuery 为核心，能兼容绝大部分浏览器，这是很多企业比较关心的，因为他们的客户很可能还在用 IE7 及以下浏览器。使用 RequireJS 或 SeaJS 进行模块化开发可以解决代码依赖混乱的问题，同时便于维护及团队协作。使用轻量级的前端模板（如 doT）可以将数据与 HTML 模板分离。最后，使用自动化构建工具（如 Gulp）可以合并压缩代码，如果你喜欢写 Less、Sass 以及现在流行的 ES 6，也可以帮你进行预编译。

这样一套看似完美无瑕的前端解决方案就构成了我们所说的传统前端开发模式，由于它的简单、高效、实用，至今仍有不少开发者在使用。不过随着项目的扩大和时间的推移，出现了更复杂的业务场景，比如 SPA（单页面富应用）、组件解耦等。为了提升开发效率，降低维护成本，传统的前端开发模式已不能完全满足我们的需求，这时就出现了如 Angular、React 以及我们要介绍的主角 Vue.js。

1.2.2　Vue.js 的开发模式

Vue.js 是一个渐进式的 JavaScript 框架，根据项目需求，你可以选择从不同的维度来使用它。如果你只是想体验 Vue.js 带来的快感，或者开发几个简单的 HTML 5 页面或小应用，你可以直接通过 script 加载 CDN 文件，例如：

```
<!-- 自动识别最新稳定版本的 Vue.js -->
<script src="https://unpkg.com/vue/dist/vue.min.js"></script>
<!--指定某个具体版本的 Vue.js -->
<script src="https://unpkg.com/vue@2.1.6/dist/vue.min.js"></script>
```

两种版本都可以，如果你不太了解各版本的差别，建议直接使用最新的稳定版本。当然，你也可以将代码下载下来，通过自己的相对路径来引用。引入 Vue.js 框架后，在 body 底部使用 new Vue()的方式创建一个实例，这就是 Vue.js 最基本的开发模式。现在可以写入以下完整的代码来快速体验 Vue：

```
<!DOCTYPE html>
<html>
<head>
    <meta charset="utf-8">
    <title>Vue 示例</title>
</head>
<body>
    <div id="app">
        <ul>
            <li v-for="book in books">{{ book.name }}</li>
        </ul>
    </div>
    <script src="https://unpkg.com/vue/dist/vue.min.js"></script>
    <script>
        new Vue({
            el: '#app',
            data: {
                books: [
                    { name: '《Vue.js 实战》' },
                    { name: '《JavaScript 语言精粹》' },
                    { name: '《JavaScript 高级程序设计》' }
                ]
            }
        })
    </script>
</body>
</html>
```

在浏览器中访问它，会将图书列表循环显示出来，如图 1-2 所示。

对于一些业务逻辑复杂，对前端工程有要求的项目，可以使用 Vue 单文件的形式配合 webpack 使用，必要时还会用到 vuex 来管理状态，vue-router 来管理路由。这里提到了很多概念，目前还不必去过多了解，只是说明 Vue.js 框架的开发模式多样化，后续章节会详细介绍，到时就会对整个 Vue 生态有所了解了。

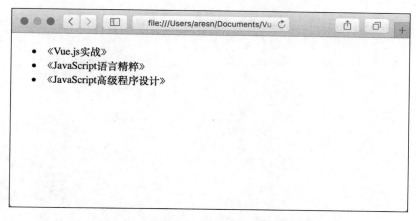

图 1-2　Vue.js 示例在浏览器中访问的效果

了解了 Vue.js 的开发模式后，相信你已经迫不及待地想开启 Vue 的大门了。下一章，我们就直接进入话题，创建第一个 Vue 应用。

第 2 章

数据绑定和第一个 Vue 应用

学习任何一种框架，从一个 Hello World 应用开始是最快了解该框架特性的途径，我们先从一段简单的 HTML 代码开始，感受 Vue.js 最核心的功能。

```html
<!DOCTYPE html>
<html>
<head>
    <meta charset="utf-8">
    <title>Vue 示例</title>
</head>
<body>
    <div id="app">
        <input type="text" v-model="name" placeholder="你的名字">
        <h1>你好，{{ name }}</h1>
    </div>
    <script src="https://unpkg.com/vue/dist/vue.min.js"></script>
    <script>
        var app = new Vue({
            el: '#app',
            data: {
                name: ''
            }
        })
    </script>
</body>
</html>
```

这是一段简单到不能再简单的代码，但却展示出了 Vue.js 最核心的功能：数据的双向绑定。在输入框输入的内容会实时展示在页面的 h1 标签内，如图 2-1 所示。

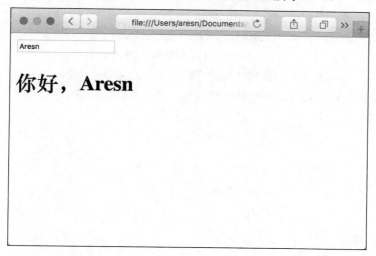

图 2-1 展示内容

提 示

从本章开始，示例不再提供完整的代码，而是根据上下文，将 HTML 部分与 JavaScript 部分单独展示，省略了 head、body 等标签以及 Vue.js 的加载等，读者可根据上例结构来组织代码。

2.1 Vue 实例与数据绑定

2.1.1 实例与数据

Vue.js 应用的创建很简单，通过构造函数 Vue 就可以创建一个 Vue 的根实例，并启动 Vue 应用：

```
var app = new Vue({
    //选项
})
```

变量 app 就代表了这个 Vue 实例。事实上，几乎所有的代码都是一个对象，写入 Vue 实例的选项内的。

首先，必不可少的一个选项就是 el。el 用于指定一个页面中已存在的 DOM 元素来挂载 Vue 实例，它可以是 HTMLElement，也可以是 CSS 选择器，比如：

```
<div id="app"></div>

var app = new Vue({
    el: document.getElementById('app')   //或者是'#app'
})
```

挂载成功后，我们可以通过 app.$el 来访问该元素。Vue 提供了很多常用的实例属性与方法，都以$开头，比如$el，后续还会介绍更多有用的方法。

回顾章节开始的 Hello World 代码，在 input 标签上，有一个 v-model 的指令，它的值对应于我们创建的 Vue 实例的 data 选项中的 name 字段，这就是 Vue 的数据绑定。

通过 Vue 实例的 data 选项，可以声明应用内需要双向绑定的数据。建议所有会用到的数据都预先在 data 内声明，这样不至于将数据散落在业务逻辑中，难以维护。

Vue 实例本身也代理了 data 对象里的所有属性，所以可以这样访问：

```
var app = new Vue({
    el: '#app',
    data: {
        a: 2
    }
})
console.log(app.a);  // 2
```

除了显式地声明数据外，也可以指向一个已有的变量，并且它们之间默认建立了双向绑定，当修改其中任意一个时，另一个也会一起变化：

```
var myData = {
    a: 1
}
var app = new Vue({
    el: '#app',
    data: myData
})
console.log(app.a);  // 1
//修改属性，原数据也会随之修改
app.a = 2;
console.log(myData.a);  // 2

// 反之，修改原数据，Vue 属性也会修改
myData.a = 3;
console.log(app.a);  // 3
```

2.1.2 生命周期

每个 Vue 实例创建时，都会经历一系列的初始化过程，同时也会调用相应的生命周期钩子，我们可以利用这些钩子，在合适的时机执行我们的业务逻辑。如果你使用过 jQuery，一定知道它的 ready()方法，比如以下示例：

```
$(document).ready(function() {
    // DOM 加载完后，会执行这里的代码
});
```

Vue 的生命周期钩子与之类似，比较常用的有：

- created 实例创建完成后调用，此阶段完成了数据的观测等，但尚未挂载，$el 还不可用。需要初始化处理一些数据时会比较有用，后面章节将有介绍。
- mounted el 挂载到实例上后调用，一般我们的第一个业务逻辑会在这里开始。
- beforeDestroy 实例销毁之前调用。主要解绑一些使用 addEventListener 监听的事件等。

这些钩子与 el 和 data 类似，也是作为选项写入 Vue 实例内，并且钩子的 this 指向的是调用它的 Vue 实例：

```
var app = new Vue({
    el: '#app',
    data: {
        a: 2
    },
    created: function () {
        console.log(this.a);  // 2
    },
    mounted: function () {
        console.log(this.$el);  // <div id="app"></div>
    }
})
```

2.1.3 插值与表达式

使用双大括号（Mustache 语法）"{{ }}"是最基本的文本插值方法，它会自动将我们双向绑定的数据实时显示出来，例如：

```
<div id="app">
    {{ book }}
</div>
<script>
    var app = new Vue({
        el: '#app',
        data: {
            book: '《Vue.js 实战》'
        }
    })
</script>
```

大括号里的内容会被替换为《Vue.js 实战》，通过任何方法修改数据 book，大括号的内容都会被实时替换，比如下面的这个示例，实时显示当前的时间，每秒更新：

```
<div id="app">
    {{ date }}
</div>
```

```
<script>
    var app = new Vue({
        el: '#app',
        data: {
            date: new Date()
        },
        mounted: function () {
            var _this = this;   // 声明一个变量指向 Vue 实例 this，保证作用域一致
            this.timer = setInterval(function () {
                _this.date = new Date();   //修改数据 date
            }, 1000);
        },
        beforeDestroy: function () {
            if (this.timer) {
                clearInterval(this.timer);   //在 Vue 实例销毁前，清除我们的定时器
            }
        }
    })
</script>
```

这里的 {{ date }} 输出的是浏览器默认的时间格式，比如 2017-01-02T14:04:49.470Z，并非格式化的时间（2017-01-02 22:04:49），所以需要注意时区。有多种方法可以对时间格式化，比如赋值前先使用自定义的函数处理。Vue 的过滤器（filter）或计算属性（computed）也可以实现，稍后会介绍到。

如果有的时候就是想输出 HTML，而不是将数据解释后的纯文本，可以使用 v-html：

```
<div id="app">
    <span v-html="link"></span>
</div>
<script>
    var app = new Vue({
        el: '#app',
        data: {
            link: '<a href="#">这是一个连接</a>'
        }
    })
</script>
```

link 的内容将会被渲染为一个具有点击功能的 a 标签，而不是纯文本。这里要注意，如果将用户产生的内容使用 v-html 输出后，有可能导致 XSS 攻击，所以要在服务端对用户提交的内容进行处理，一般可将尖括号 "<>" 转义。

如果想显示{{}}标签,而不进行替换,使用 v-pre 即可跳过这个元素和它的子元素的编译过程,例如:

```
<span v-pre>{{ 这里的内容是不会被编译的}}</span>
```

在{{}}中,除了简单的绑定属性值外,还可以使用 JavaScript 表达式进行简单的运算、三元运算等,例如:

```
<div id="app">
    {{ number / 10 }}
    {{ isOK ? '确定' : '取消' }}
    {{ text.split(',').reverse().join(',') }}
</div>
<script>
    var app = new Vue({
        el: '#app',
        data: {
            number: 100,
            isOK: false,
            text: '123,456'
        }
    })
</script>
```

显示结果依次为:10、取消、456,123。

Vue.js 只支持单个表达式,不支持语句和流控制。另外,在表达式中,不能使用用户自定义的全局变量,只能使用 Vue 白名单内的全局变量,例如 Math 和 Date。以下是一些无效的示例:

```
<!-- 这是语句,不是表达式 -->
{{ var book = 'Vue.js 实战' }}
<!--不能使用流控制,要使用三元运算 -->
{{ if (ok) return msg }}
```

2.1.4　过滤器

Vue.js 支持在{{}}插值的尾部添加一个管道符"(|)"对数据进行过滤,经常用于格式化文本,比如字母全部大写、货币千位使用逗号分隔等。过滤的规则是自定义的,通过给 Vue 实例添加选项 filters 来设置,例如在上一节中实时显示当前时间的示例,可以对时间进行格式化处理:

```
<div id="app">
    {{ date | formatDate }}
</div>
<script>
    //在月份、日期、小时等小于 10 时前面补 0
    var padDate = function(value) {
        return value < 10 ? '0' + value : value;
```

```
        };

        var app = new Vue({
            el: '#app',
            data: {
                date: new Date()
            },
            filters: {
                formatDate: function (value) {   // 这里的value 就是需要过滤的数据
                    var date = new Date(value);
                    var year = date.getFullYear();
                    var month = padDate(date.getMonth() + 1);
                    var day = padDate(date.getDate());
                    var hours = padDate(date.getHours());
                    var minutes = padDate(date.getMinutes());
                    var seconds = padDate(date.getSeconds());
                    //将整理好的数据返回出去
                    return year + '-' + month + '-' + day + ' ' + hours + ':' + minutes + ':' + seconds;
                }
            },
            mounted: function () {
                var _this = this;  // 声明一个变量指向 Vue 实例 this，保证作用域一致
                this.timer = setInterval(function () {
                    _this.date = new Date();   //修改数据 date
                }, 1000);
            },
            beforeDestroy: function () {
                if (this.timer) {
                    clearInterval(this.timer);   //在 Vue 实例销毁前，清除我们的定时器
                }
            }
        })
</script>
```

过滤器也可以串联，而且可以接收参数，例如：

```
<!-- 串联 -->
{{ message | filterA | filterB }}

<!-- 接收参数-->
{{ message | filterA('arg1', 'arg2') }}
```

这里的字符串 arg1 和 arg2 将分别传给过滤器的第二个和第三个参数，因为第一个是数据本身。

过滤器应当用于处理简单的文本转换，如果要实现更为复杂的数据变换，应该使用计算属性，下一章中会详细介绍它的用法。

2.2 指令与事件

指令（Directives）是 Vue.js 模板中最常用的一项功能，它带有前缀 v-，在前文我们已经使用过不少指令了，比如 v-if、v-html、v-pre 等。指令的主要职责就是当其表达式的值改变时，相应地将某些行为应用到 DOM 上，以 v-if 为例：

```
<div id="app">
    <p v-if="show">显示这段文本</p>
</div>
<script>
    var app = new Vue({
        el: '#app',
        data: {
            show: true
        }
    })
</script>
```

当数据 show 的值为 true 时，p 元素会被插入，为 false 时则会被移除。数据驱动 DOM 是 Vue.js 的核心理念，所以不到万不得已时不要主动操作 DOM，你只需要维护好数据，DOM 的事 Vue 会帮你优雅的处理。

Vue.js 内置了很多指令，帮助我们快速完成常见的 DOM 操作，比如循环渲染、显示与隐藏等。在第 5 章会详细地介绍这些内置指令，但在此之前，你需要先知道 v-bind 和 v-on。

v-bind 的基本用途是动态更新 HTML 元素上的属性，比如 id、class 等，例如下面几个示例：

```
<div id="app">
    <a v-bind:href="url">链接</a>
    <img v-bind:src="imgUrl">
</div>
<script>
    var app = new Vue({
        el: '#app',
        data: {
            url: 'https://www.github.com',
            imgUrl: 'http://xxx.xxx.xx/img.png'
        }
    })
</script>
```

示例中的链接地址与图片的地址都与数据进行了绑定，当通过各种方式改变数据时，链接和图片都会自动更新。上述示例渲染后的结果为：

```
<a href="https://www.github.com">链接</a>
<img src="http://xxx.xxx.xx/img.png">
```

以上是介绍 v-bind 最基本的用法，它在 Vue.js 组件中还有着重要的作用，将在第 4 章和第 7 章中详细介绍。

另一个非常重要的指令就是 v-on，它用来绑定事件监听器，这样我们就可以做一些交互了，先来看下面的示例：

```
<div id="app">
    <p v-if="show">这是一段文本</p>
    <button v-on:click="handleClose">点击隐藏</button>
</div>
<script>
    var app = new Vue({
        el: '#app',
        data: {
            show: true
        },
        methods: {
            handleClose: function () {
                this.show = false;
            }
        }
    })
</script>
```

在 button 按钮上，使用 v-on:click 给该元素绑定了一个点击事件，在普通元素上，v-on 可以监听原生的 DOM 事件，除了 click 外，还有 dblclick、keyup、mousemove 等。表达式可以是一个方法名，这些方法都写在 Vue 实例的 methods 属性内，并且是函数的形式，函数内的 this 指向的是当前 Vue 实例本身，因此可以直接使用 this.xxx 的形式来访问或修改数据，如示例中的 this.show = false; 把数据 show 修改为了 false，所以点击按钮时，文本 p 元素就被移除了。

表达式除了方法名，也可以直接是一个内联语句，上例也可以改写为：

```
<div id="app">
    <p v-if="show">这是一段文本</p>
    <button v-on:click="show = false">点击隐藏</button>
</div>
<script>
    var app = new Vue({
        el: '#app',
        data: {
            show: true
```

```
        }
    })
</script>
```

如果绑定的事件要处理复杂的业务逻辑，建议还是在 methods 里声明一个方法，这样可读性更强也好维护。

Vue.js 将 methods 里的方法也代理了，所以也可以像访问 Vue 数据那样来调用方法：

```
<div id="app">
    <p v-if="show">这是一段文本</p>
    <button v-on:click="handleClose">点击隐藏</button>
</div>
<script>
    var app = new Vue({
        el: '#app',
        data: {
            show: true
        },
        methods: {
            handleClose: function () {
                this.close();
            },
            close: function () {
                this.show = false;
            }
        }
    })
</script>
```

在 handleClose 方法内，直接通过 this.close() 调用了 close() 函数。在上面示例中是多此一举的，只是用于演示它的用法，在业务中会经常用到，例如以下几种用法都是正确的：

```
<script>
    var app = new Vue({
        el: '#app',
        data: {
            show: true
        },
        methods: {
            init: function (text) {
                console.log(text);
            }
        },
        mounted: function () {
            this.init('在初始化时调用');    //在初始化时调用
```

```
        }
    });

    app.init('通过外部调用');    //在 Vue 实例外部调用
</script>
```

更多关于 v-on 事件的用法将会在第 7 章中详细介绍。

2.3 语 法 糖

语法糖是指在不影响功能的情况下，添加某种方法实现同样的效果，从而方便程序开发。

Vue.js 的 v-bind 和 v-on 指令都提供了语法糖，也可以说是缩写，比如 v-bind，可以省略 v-bind，直接写一个冒号 ":"：

```
<a v-bind:href="url">链接</a>
<img v-bind:src="imgUrl">
<!-- 缩写为-->
<a :href="url">链接</a>
<img :src="imgUrl">
```

v-on 可以直接用 "@" 来缩写：

```
<button v-on:click="handleClose">点击隐藏</button>
<!-- 缩写为-->
<button @click="handleClose">点击隐藏</button>
```

使用语法糖可以简化代码的书写，从下一章开始，所有示例的 v-bind 和 v-on 指令将默认使用语法糖的写法。

第 3 章

计算属性

模板内的表达式常用于简单的运算,当其过长或逻辑复杂时,会难以维护,本章的计算属性就是用于解决该问题的。

3.1 什么是计算属性

通过上一章的介绍,我们已经可以搭建出一个简单的 Vue 应用,在模板中双向绑定一些数据或表达式了。但是表达式如果过长,或逻辑更为复杂时,就会变得臃肿甚至难以阅读和维护,比如:

```
<div>
    {{ text.split(',').reverse().join(',') }}
</div>
```

这里的表达式包含 3 个操作,并不是很清晰,所以在遇到复杂的逻辑时应该使用计算属性。上例可以用计算属性进行改写:

```
<div id="app">
    {{ reversedText }}
</div>
<script>
    var app = new Vue({
        el: '#app',
        data: {
            text: '123,456'
```

```
        },
        computed: {
            reversedText: function () {
                // 这里的 this 指向的是当前的 Vue 实例
                return this.text.split(',').reverse().join(',');
            }
        }
    })
</script>
```

所有的计算属性都以函数的形式写在 Vue 实例内的 computed 选项内，最终返回计算后的结果。

3.2 计算属性用法

在一个计算属性里可以完成各种复杂的逻辑，包括运算、函数调用等，只要最终返回一个结果就可以。除了上例简单的用法，计算属性还可以依赖多个 Vue 实例的数据，只要其中任一数据变化，计算属性就会重新执行，视图也会更新。例如，下面的示例展示的是在购物车内两个包裹的物品总价：

```
<div id="app">
    总价: {{ prices }}
</div>
<script>
    var app = new Vue({
        el: '#app',
        data: {
            package1: [
                {
                    name: 'iPhone 7',
                    price: 7199,
                    count: 2
                },
                {
                    name: 'iPad',
                    price: 2888,
                    count: 1
                }
            ],
            package2: [
                {
                    name: 'apple',
                    price: 3,
```

```
                    count: 5
                },
                {
                    name: 'banana',
                    price: 2,
                    count: 10
                }
            ]
        },
        computed: {
            prices: function () {
                var prices = 0;
                for (var i = 0; i < this.package1.length; i++) {
                    prices += this.package1[i].price * this.package1[i].count;
                }
                for (var i = 0; i < this.package2.length; i++) {
                    prices += this.package2[i].price * this.package2[i].count;
                }
                return prices;
            }
        }
    })
</script>
```

当 package1 或 package2 中的商品有任何变化，比如购买数量变化或增删商品时，计算属性 prices 就会自动更新，视图中的总价也会自动变化。

每一个计算属性都包含一个 getter 和一个 setter，我们上面的两个示例都是计算属性的默认用法，只是利用了 getter 来读取。在你需要时，也可以提供一个 setter 函数，当手动修改计算属性的值就像修改一个普通数据那样时，就会触发 setter 函数，执行一些自定义的操作，例如：

```
<div id="app">
    姓名：{{ fullName }}
</div>
<script>
var app = new Vue({
    el: '#app',
    data: {
        firstName: 'Jack',
        lastName: 'Green'
    },
    computed: {
        fullName: {
            // getter，用于读取
```

```
            get: function () {
                return this.firstName + ' ' + this.lastName;
            },
            // setter,写入时触发
            set: function (newValue) {
                var names = newValue.split(' ');
                this.firstName = names[0];
                this.lastName = names[names.length - 1];
            }
        }
    }
});
</script>
```

当执行 app.fullName = 'John Doe';时,setter 就会被调用,数据 firstName 和 lastName 都会相对更新,视图同样也会更新。

绝大多数情况下,我们只会用默认的 getter 方法来读取一个计算属性,在业务中很少用到 setter,所以在声明一个计算属性时,可以直接使用默认的写法,不必将 getter 和 setter 都声明。

计算属性除了上述简单的文本插值外,还经常用于动态地设置元素的样式名称 class 和内联样式 style,在下章会介绍这方面的内容。当使用组件时,计算属性也经常用来动态传递 props,这也会在第 7 章组件里详细介绍。

计算属性还有两个很实用的小技巧容易被忽略:一是计算属性可以依赖其他计算属性;二是计算属性不仅可以依赖当前 Vue 实例的数据,还可以依赖其他实例的数据,例如:

```
<div id="app1"></div>
<div id="app2">
    {{ reversedText }}
</div>
<script>
    var app1 = new Vue({
        el: '#app1',
        data: {
            text: '123,456'
        }
    });

    var app2 = new Vue({
        el: '#app2',
        computed: {
            reversedText: function () {
                // 这里依赖的是实例 app1 的数据 text
```

```
            return app1.text.split(',').reverse().join(',');
        }
    }
})
</script>
```

这里我们创建了两个 Vue 实例 app1 和 app2,在 app2 的计算属性 reversedText 中,依赖的是 app1 的数据 text,所以当 text 变化时,实例 app2 的计算属性也会变化。这样的用法在后面章节介绍的组件和组件化里会用到,尤其是在多人协同开发时很常用,因为你写的一个组件所用得到的数据需要依赖他人的组件提供。随着后面对组件的深入会慢慢意识到这点,现在可以不用太过了解。

3.3 计算属性缓存

在上一章介绍指令与事件时,你可能发现调用 methods 里的方法也可以与计算属性起到同样的作用,比如本章第一个示例可以用 methods 改写为:

```
<div id="app">
    <!-- 注意,这里的 reversedText 是方法,所以要带() -->
    {{ reversedText() }}
</div>
<script>
    var app = new Vue({
        el: '#app',
        data: {
            text: '123,456'
        },
        methods: {
            reversedText: function () {
                // 这里的 this 指向的是当前 Vue 实例
                return this.text.split(',').reverse().join(',');
            }
        }
    })
</script>
```

没有使用计算属性,在 methods 里定义了一个方法实现了相同的效果,甚至该方法还可以接受参数,使用起来更灵活。既然使用 methods 就可以实现,那么为什么还需要计算属性呢?原因就是计算属性是基于它的依赖缓存的。一个计算属性所依赖的数据发生变化时,它才会重新取值,所以 text 只要不改变,计算属性也就不更新,例如:

```
computed: {
    now: function () {
```

```
        return Date.now();
    }
}
```

这里的 Date.now() 不是响应式依赖，所以计算属性 now 不会更新。但是 methods 则不同，只要重新渲染，它就会被调用，因此函数也会被执行。

使用计算属性还是 methods 取决于你是否需要缓存，当遍历大数组和做大量计算时，应当使用计算属性，除非你不希望得到缓存。

第 4 章

v-bind 及 class 与 style 绑定

DOM 元素经常会动态地绑定一些 class 类名或 style 样式，本章将介绍使用 v-bind 指令来绑定 class 和 style 的多种方法。

4.1 了解 v-bind 指令

在第 2 章时，我们已经介绍了指令 v-bind 的基本用法以及它的语法糖，它的主要用法是动态更新 HTML 元素上的属性，回顾一下下面的示例：

```
<div id="app">
    <a v-bind:href="url">链接</a>
    <img v-bind:src="imgUrl">
    <!-- 缩写为-->
    <a :href="url">链接</a>
    <img :src="imgUrl">
</div>
<script>
    var app = new Vue({
        el: '#app',
        data: {
            url: 'https://www.github.com',
            imgUrl: 'http://xxx.xxx.xx/img.png'
        }
    })
</script>
```

链接的 href 属性和图片的 src 属性都被动态设置了,当数据变化时,就会重新渲染。

在数据绑定中,最常见的两个需求就是元素的样式名称 class 和内联样式 style 的动态绑定,它们也是 HTML 的属性,因此可以使用 v-bind 指令。我们只需要用 v-bind 计算出表达式最终的字符串就可以,不过有时候表达式的逻辑较复杂,使用字符串拼接方法较难阅读和维护,所以 Vue.js 增强了对 class 和 style 的绑定。

4.2 绑定 class 的几种方式

4.2.1 对象语法

给 v-bind:class 设置一个对象,可以动态地切换 class,例如:

```
<div id="app">
    <div :class="{ 'active': isActive }"></div>
</div>
<script>
    var app = new Vue({
        el: '#app',
        data: {
            isActive: true
        }
    })
</script>
```

上面的 :class 等同于 v-bind:class,是一个语法糖,如不特殊说明,后面都将使用语法糖写法,可以回顾第 2.3 节。

上面示例中,类名 active 依赖于数据 isActive,当其为 true 时,div 会拥有类名 Active,为 false 时则没有,所以上例最终渲染完的结果是:

```
<div class="active"></div>
```

对象中也可以传入多个属性,来动态切换 class。另外,:class 可以与普通 class 共存,例如:

```
<div id="app">
    <div class="static" :class="{ 'active': isActive, 'error': isError }"></div>
</div>
<script>
    var app = new Vue({
        el: '#app',
        data: {
            isActive: true,
```

```
            isError: false
        }
    })
</script>
```

:class 内的表达式每项为真时，对应的类名就会加载，上面渲染后的结果为：

```
<div class="static active"></div>
```

当数据 isActive 或 isError 变化时，对应的 class 类名也会更新。比如当 isError 为 true 时，渲染后的结果为：

```
<div class="static active error"></div>
```

当 :class 的表达式过长或逻辑复杂时，还可以绑定一个计算属性，这是一种很友好和常见的用法，一般当条件多于两个时，都可以使用 data 或 computed，例如使用计算属性：

```
<div id="app">
    <div :class="classes"></div>
</div>
<script>
    var app = new Vue({
        el: '#app',
        data: {
            isActive: true,
            error: null
        },
        computed: {
            classes: function () {
                return {
                    active: this.isActive && !this.error,
                    'text-fail': this.error && this.error.type === 'fail'
                }
            }
        }
    })
</script>
```

除了计算属性，你也可以直接绑定一个 Object 类型的数据，或者使用类似计算属性的 methods。

4.2.2 数组语法

当需要应用多个 class 时，可以使用数组语法，给 :class 绑定一个数组，应用一个 class 列表：

```
<div id="app">
    <div :class="[activeCls, errorCls]"></div>
</div>
```

```
<script>
    var app = new Vue({
        el: '#app',
        data: {
            activeCls: 'active',
            errorCls: 'error'
        }
    })
</script>
```

渲染后的结果为：

```
<div class="active error"></div>
```

也可以使用三元表达式来根据条件切换 class，例如下面的示例：

```
<div id="app">
    <div :class="[isActive ? activeCls : '', errorCls]"></div>
</div>
<script>
    var app = new Vue({
        el: '#app',
        data: {
            isActive: true,
            activeCls: 'active',
            errorCls: 'error'
        }
    })
</script>
```

样式 error 会始终应用，当数据 isActive 为真时，样式 active 才会被应用。class 有多个条件时，这样写较为烦琐，可以在数组语法中使用对象语法：

```
<div id="app">
    <div :class="[{ 'active': isActive }, errorCls]"></div>
</div>
<script>
    var app = new Vue({
        el: '#app',
        data: {
            isActive: true,
            errorCls: 'error'
        }
    })
</script>
```

当然，与对象语法一样，也可以使用 data、computed 和 methods 三种方法，以计算属性为例：

```
<div id="app">
    <button :class="classes"></button>
</div>
<script>
    var app = new Vue({
        el: '#app',
        data: {
            size: 'large',
            disabled: true
        },
        computed: {
            classes: function () {
                return [
                    'btn',
                    {
                        ['btn-' + this.size]: this.size !== '',
                        ['btn-disabled']: this.disabled
                    }
                ];
            }
        }
    })
</script>
```

示例中的样式 btn 会始终应用，当数据 size 不为空时，会应用样式前缀 btn-，后加 size 的值；当数据 disabled 为真时，会应用样式 btn-disabled，所以该示例最终渲染的结果为：

```
<button class="btn btn-large btn-disabled"></button>
```

使用计算属性给元素动态设置类名，在业务中经常用到，尤其是在写复用的组件时，所以在开发过程中，如果表达式较长或逻辑复杂，应该尽可能地优先使用计算属性。

4.2.3 在组件上使用

提示

本节内容依赖第 7 章组件相关的内容，如果你尚未了解过 Vue.js 的组件，可以先跳过这节，稍后再阅读。

如果直接在自定义组件上使用 class 或 :class，样式规则会直接应用到这个组件的根元素上，例如声明一个简单的组件：

```
Vue.component('my-component', {
    template: '<p class="article">一些文本</p>'
});
```

然后在调用这个组件时，应用上面两节介绍的对象语法或数组语法给组件绑定 class，以对象语法为例：

```
<div id="app">
    <my-component :class="{ 'active': isActive }"></my-component>
</div>
<script>
    var app = new Vue({
        el: '#app',
        data: {
            isActive: true
        }
    })
</script>
```

最终组件渲染后的结果为：

```
<p class="article active">一些文本</p>
```

这种用法仅适用于自定义组件的最外层是一个根元素，否则会无效，当不满足这种条件或需要给具体的子元素设置类名时，应当使用组件的 props 来传递。这些用法同样适用于下一节中绑定内联样式 style 的内容。

4.3 绑定内联样式

使用 v-bind:style（即 :style）可以给元素绑定内联样式，方法与:class 类似，也有对象语法和数组语法，看起来很像直接在元素上写 CSS：

```
<div id="app">
    <div :style="{ 'color': color, 'fontSize': fontSize + 'px' }">文本</div>
</div>
<script>
    var app = new Vue({
        el: '#app',
        data: {
            color: 'red',
            fontSize: 14
        }
    })
</script>
```

CSS 属性名称使用驼峰命名（camelCase）或短横分隔命名（kebab-case），渲染后的结果为：

```
<div style="color: red; font-size: 14px;">文本</div>
```

大多数情况下，直接写一长串的样式不便于阅读和维护，所以一般写在 data 或 computed 里，以 data 为例改写上面的示例：

```
<div id="app">
    <div :style="styles">文本</div>
</div>
<script>
    var app = new Vue({
        el: '#app',
        data: {
            styles: {
                color: 'red',
                fontSize: 14 + 'px'
            }
        }
    })
</script>
```

应用多个样式对象时，可以使用数组语法：

```
<div :style="[styleA, styleB]">文本</div>
```

在实际业务中，:style 的数组语法并不常用，因为往往可以写在一个对象里面；而较为常用的应当是计算属性。

另外，使用:style 时，Vue.js 会自动给特殊的 CSS 属性名称增加前缀，比如 transform。

第 5 章

内 置 指 令

回顾一下第 2.2 节，我们已经介绍过指令（Directive）的概念了，Vue.js 的指令是带有特殊前缀 "v-" 的 HTML 特性，它绑定一个表达式，并将一些特性应用到 DOM 上。其实我们已经用到过很多 Vue 内置的指令，比如 v-html、v-pre，还有上一章的 v-bind。本章将继续介绍 Vue.js 中更多常用的内置指令。

5.1 基本指令

5.1.1 v-cloak

v-cloak 不需要表达式，它会在 Vue 实例结束编译时从绑定的 HTML 元素上移除，经常和 CSS 的 display: none; 配合使用：

```
<div id="app" v-cloak>
    {{ message }}
</div>
<script>
    var app = new Vue({
        el: '#app',
        data: {
            message: '这是一段文本'
        }
    })
</script>
```

这时虽然已经加了指令 v-cloak，但其实并没有起到任何作用，当网速较慢、Vue.js 文件还没加载完时，在页面上会显示{{ message }}的字样，直到 Vue 创建实例、编译模板时，DOM 才会被替换，所以这个过程屏幕是有闪动的。只要加一句 CSS 就可以解决这个问题了：

```css
[v-cloak] {
    display: none;
}
```

在一般情况下，v-cloak 是一个解决初始化慢导致页面闪动的最佳实践，对于简单的项目很实用，但是在具有工程化的项目里，比如后面进阶篇将介绍 webpack 和 vue-router 时，项目的 HTML 结构只有一个空的 div 元素，剩余的内容都是由路由去挂载不同组件完成的，所以不再需要 v-cloak。

5.1.2 v-once

v-once 也是一个不需要表达式的指令，作用是定义它的元素或组件只渲染一次，包括元素或组件的所有子节点。首次渲染后，不再随数据的变化重新渲染，将被视为静态内容，例如：

```html
<div id="app">
    <span v-once>{{ message }}</div>
    <div v-once>
        <span>{{ message }}</span>
    </div>
</div>
<script>
    var app = new Vue({
        el: '#app',
        data: {
            message: '这是一段文本'
        }
    })
</script>
```

v-once 在业务中也很少使用，当你需要进一步优化性能时，可能会用到。

5.2 条件渲染指令

5.2.1 v-if、v-else-if、v-else

与 JavaScript 的条件语句 if、else、else if 类似，Vue.js 的条件指令可以根据表达式的值在 DOM 中渲染或销毁元素/组件，例如：

```html
<div id="app">
    <p v-if="status === 1">当 status 为 1 时显示该行</p>
    <p v-else-if="status === 2">当 status 为 2 时显示该行</p>
    <p v-else>否则显示该行</p>
```

```
    </div>
    <script>
        var app = new Vue({
            el: '#app',
            data: {
                status: 1
            }
        })
    </script>
```

v-else-if 要紧跟 **v-if**，**v-else** 要紧跟 **v-else-if** 或 **v-if**，表达式的值为真时，当前元素/组件及所有子节点将被渲染，为假时被移除。如果一次判断的是多个元素，可以在 Vue.js 内置的 <template> 元素上使用条件指令，最终渲染的结果不会包含该元素，例如：

```
    <div id="app">
        <template v-if="status === 1">
            <p>这是一段文本</p>
            <p>这是一段文本</p>
            <p>这是一段文本</p>
        </template>
    </div>
    <script>
        var app = new Vue({
            el: '#app',
            data: {
                status: 1
            }
        })
    </script>
```

Vue 在渲染元素时，出于效率考虑，会尽可能地复用已有的元素而非重新渲染，比如下面的示例：

```
    <div id="app">
        <template v-if="type === 'name'">
            <label>用户名：</label>
            <input placeholder="输入用户名">
        </template>
        <template v-else>
            <label>邮箱：</label>
            <input placeholder="输入邮箱">
        </template>
        <button @click="handleToggleClick">切换输入类型</button>
    </div>
    <script>
```

```
    var app = new Vue({
      el: '#app',
      data: {
        type: 'name'
      },
      methods: {
        handleToggleClick: function() {
          this.type = this.type === 'name' ? 'mail' : 'name';
        }
      }
    })
</script>
```

如图 5-1 和图 5-2 所示，键入内容后，点击切换按钮，虽然 DOM 变了，但是之前在输入框键入的内容并没有改变，只是替换了 placeholder 的内容，说明<input>元素被复用了。

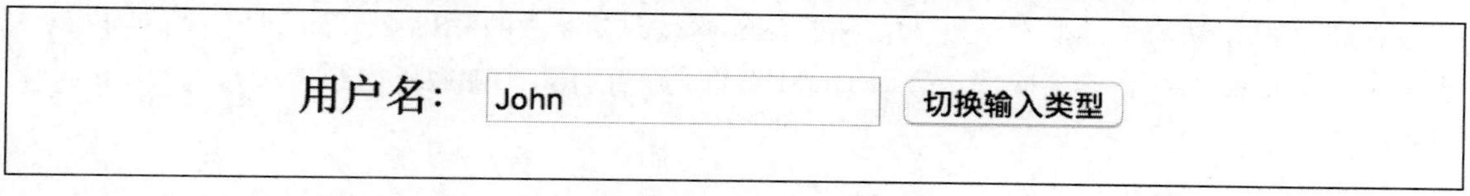

图 5-1 切换前的状态

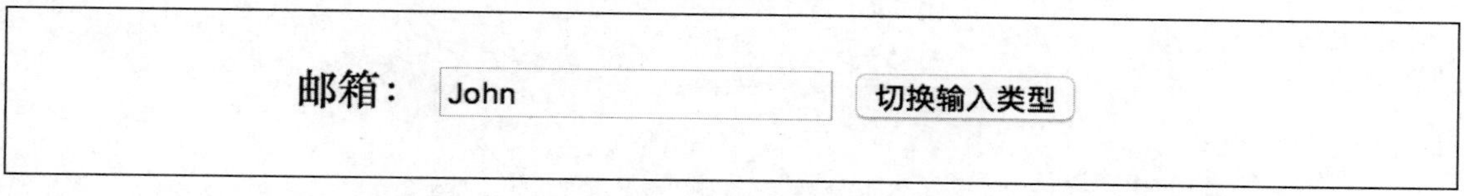

图 5-2 切换后的状态

如果你不希望这样做，可以使用 Vue.js 提供的 key 属性，它可以让你自己决定是否要复用元素，key 的值必须是唯一的，例如：

```
<div id="app">
    <template v-if="type === 'name'">
        <label>用户名：</label>
        <input placeholder="输入用户名" key="name-input">
    </template>
    <template v-else>
        <label>邮箱：</label>
        <input placeholder="输入邮箱" key="mail-input">
    </template>
    <button @click="handleToggleClick">切换输入类型</button>
</div>
<script>
    var app = new Vue({
        el: '#app',
```

```
        data: {
            type: 'name'
        },
        methods: {
            handleToggleClick: function() {
                this.type = this.type === 'name' ? 'mail' : 'name';
            }
        }
    })
</script>
```

给两个<input>元素都增加 key 后,就不会复用了,切换类型时键入的内容也会被删除,不过<label>元素仍然是被复用的,因为没有添加 key 属性。

5.2.2 v-show

v-show 的用法与 v-if 基本一致,只不过 v-show 是改变元素的 CSS 属性 display。当 v-show 表达式的值为 false 时,元素会隐藏,查看 DOM 结构会看到元素上加载了内联样式 display: none;,例如:

```
<div id="app">
    <p v-show="status === 1">当 status 为 1 时显示该行</p>
</div>
<script>
    var app = new Vue({
        el: '#app',
        data: {
            status: 2
        }
    })
</script>
```

渲染后的结果为:

```
<p style="display: none;">当 status 为 1 时显示该行</p>
```

v-show 不能在<template>上使用。

5.2.3 v-if 与 v-show 的选择

v-if 和 v-show 具有类似的功能,不过 v-if 才是真正的条件渲染,它会根据表达式适当地销毁或重建元素及绑定的事件或子组件。若表达式初始值为 false,则一开始元素/组件并不会渲染,只有当条件第一次变为真时才开始编译。

而 v-show 只是简单的 CSS 属性切换，无论条件真与否，都会被编译。相比之下，v-if 更适合条件不经常改变的场景，因为它切换开销相对较大，而 v-show 适用于频繁切换条件。

5.3 列表渲染指令 v-for

5.3.1 基本用法

当需要将一个数组遍历或枚举一个对象循环显示时，就会用到列表渲染指令 v-for。它的表达式需结合 in 来使用，类似 item in items 的形式，看下面的示例：

```
<div id="app">
    <ul>
        <li v-for="book in books">{{ book.name }}</li>
    </ul>
</div>
<script>
    var app = new Vue({
        el: '#app',
        data: {
            books: [
                { name: '《Vue.js 实战》' },
                { name: '《JavaScript 语言精粹》' },
                { name: '《JavaScript 高级程序设计》' }
            ]
        }
    })
</script>
```

我们定义一个数组类型的数据 books，用 v-for 将<li>标签循环渲染，效果如图 5-3 所示。

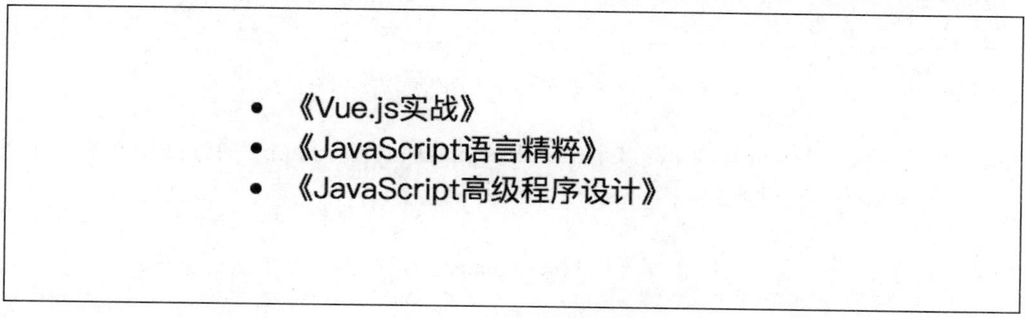

- 《Vue.js实战》
- 《JavaScript语言精粹》
- 《JavaScript高级程序设计》

图 5-3 列表循环结果

在表达式中，books 是数据，book 是当前数组元素的别名，循环出的每个<li>内的元素都可以访问到对应的当前数据 book。列表渲染也支持用 of 来代替 in 作为分隔符，它更接近 JavaScript 迭代器的语法：

```
<li v-for="book of books">{{ book.name }}</li>
```

v-for 的表达式支持一个可选参数作为当前项的索引，例如：

```
<div id="app">
    <ul>
        <li v-for="(book, index) in books">{{ index }} - {{ book.name }}</li>
    </ul>
</div>
<script>
    var app = new Vue({
        el: '#app',
        data: {
            books: [
                { name: '《Vue.js 实战》' },
                { name: '《JavaScript 语言精粹》' },
                { name: '《JavaScript 高级程序设计》' }
            ]
        }
    })
</script>
```

分隔符 in 前的语句使用括号，第二项就是 books 当前项的索引，渲染后的结果如图 5-4 所示。

- 0 – 《Vue.js实战》
- 1 – 《JavaScript语言精粹》
- 2 – 《JavaScript高级程序设计》

图 5-4　含有 index 选项的列表渲染结果

如果你使用过 Vue.js 1.x 的版本，这里的 index 也可以由内置的 $index 代替，不过在 2.x 里取消了该用法。

与 v-if 一样，v-for 也可以用在内置标签 <template> 上，将多个元素进行渲染：

```
<div id="app">
    <ul>
        <template v-for="book in books">
            <li>书名：{{ book.name }}</li>
            <li>作者：{{ book.author }}</li>
```

```
            </template>
        </ul>
</div>
<script>
    var app = new Vue({
        el: '#app',
        data: {
            books: [
                {
                    name: '《Vue.js 实战》',
                    author: '梁灏'
                },
                {
                    name: '《JavaScript 语言精粹》',
                    author: 'Douglas Crockford'
                },
                {
                    name: '《JavaScript 高级程序设计》',
                    author: 'Nicholas C.Zakas'
                }
            ]
        }
    })
</script>
```

除了数组外,对象的属性也是可以遍历的,例如:

```
<div id="app">
    <span v-for="value in user">{{ value }} </span>
</div>
<script>
    var app = new Vue({
        el: '#app',
        data: {
            user: {
                name: 'Aresn',
                gender: '男',
                age: 26
            }
        }
    })
</script>
```

渲染后的结果为：

`<span>Aresn </span><span>男 </span><span>26 </span>`

遍历对象属性时，有两个可选参数，分别是键名和索引：

```html
<div id="app">
    <ul>
        <li v-for="(value, key, index) in user">
            {{ index }} - {{ key }}: {{ value }}
        </li>
    </ul>
</div>
<script>
    var app = new Vue({
        el: '#app',
        data: {
            user: {
                name: 'Aresn',
                gender: '男',
                age: 26
            }
        }
    })
</script>
```

渲染后的结果如图 5-5 所示。

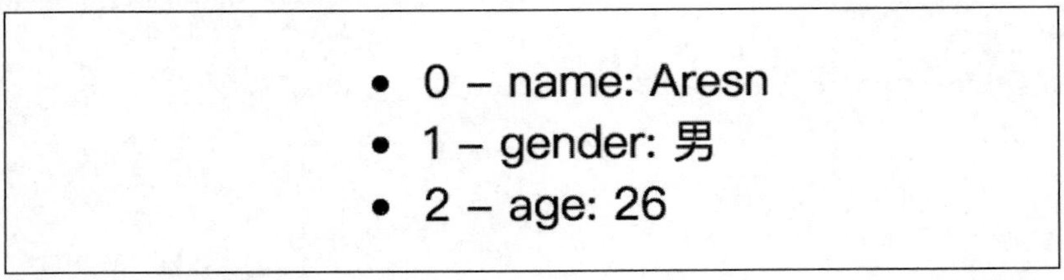

- 0 – name: Aresn
- 1 – gender: 男
- 2 – age: 26

图 5-5　遍历对象的渲染结果

v-for 还可以迭代整数：

```html
<div id="app">
    <span v-for="n in 10">{{ n }} </span>
</div>
<script>
    var app = new Vue({
        el: '#app'
    })
</script>
```

渲染后的结果为：

1 2 3 4 5 6 7 8 9 10

5.3.2 数组更新

Vue 的核心是数据与视图的双向绑定，当我们修改数组时，Vue 会检测到数据变化，所以用 v-for 渲染的视图也会立即更新。Vue 包含了一组观察数组变异的方法，使用它们改变数组也会触发视图更新：

- push()
- pop()
- shift()
- unshift()
- splice()
- sort()
- reverse()

例如，我们将之前一个示例的数据 books 添加一项：

```
app.books.push({
    name: '《CSS 揭秘》',
    author: '[希] Lea Verou'
});
```

可以尝试编写完整示例来查看效果。

使用以上方法会改变被这些方法调用的原始数组，有些方法不会改变原数组，例如：

- filter()
- concat()
- slice()

它们返回的是一个新数组，在使用这些非变异方法时，可以用新数组来替换原数组，还是之前展示书目的示例，我们找出含有 JavaScript 关键词的书目，例如：

```
<div id="app">
    <ul>
        <template v-for="book in books">
            <li>书名：{{ book.name }}</li>
            <li>作者：{{ book.author }}</li>
        </template>
    </ul>
</div>
<script>
    var app = new Vue({
        el: '#app',
```

```
        data: {
            books: [
                {
                    name: '《Vue.js 实战》',
                    author: '梁灏'
                },
                {
                    name: '《JavaScript 语言精粹》',
                    author: 'Douglas Crockford'
                },
                {
                    name: '《JavaScript 高级程序设计》',
                    author: 'Nicholas C.Zakas'
                }
            ]
        }
    });

    app.books = app.books.filter(function (item) {
        return item.name.match(/JavaScript/);
    });
</script>
```

渲染的结果中，第一项《Vue.js 实战》被过滤掉了，只显示了书名中含有 JavaScript 的选项。

Vue 在检测到数组变化时，并不是直接重新渲染整个列表，而是最大化地复用 DOM 元素。替换的数组中，含有相同元素的项不会被重新渲染，因此可以大胆地用新数组来替换旧数组，不用担心性能问题。

需要注意的是，以下变动的数组中，Vue 是不能检测到的，也不会触发视图更新：

- 通过索引直接设置项，比如 app.books[3] = {...}。
- 修改数组长度，比如 app.books.length = 1。

解决第一个问题可以用两种方法实现同样的效果，第一种是使用 Vue 内置的 set 方法：

```
Vue.set(app.books, 3, {
    name: '《CSS 揭秘》',
    author: '[希] Lea Verou'
});
```

如果是在 webpack 中使用组件化的方式（进阶篇中将介绍），默认是没有导入 Vue 的，这时可以使用$set，例如：

```
this.$set(app.books, 3, {
    name: '《CSS 揭秘》',
    author: '[希] Lea Verou'
```

```
})
```
// 这里的 this 指向的就是当前组件实例,即 app。在非 webpack 模式下也可以用$set 方法,例如 app.$set(…)

另一种方法:

```
app.books.splice(3, 1, {
    name: '《CSS 揭秘》',
    author: '[希] Lea Verou'
})
```

第二个问题也可以直接用 splice 来解决:

```
app.books.splice(1);
```

5.3.3　过滤与排序

当你不想改变原数组,想通过一个数组的副本来做过滤或排序的显示时,可以使用计算属性来返回过滤或排序后的数组,例如:

```
<div id="app">
    <ul>
        <template v-for="book in filterBooks">
            <li>书名: {{ book.name }}</li>
            <li>作者: {{ book.author }}</li>
        </template>
    </ul>
</div>
<script>
    var app = new Vue({
        el: '#app',
        data: {
            books: [
                {
                    name: '《Vue.js 实战》',
                    author: '梁灏'
                },
                {
                    name: '《JavaScript 语言精粹》',
                    author: 'Douglas Crockford'
                },
                {
                    name: '《JavaScript 高级程序设计》',
                    author: 'Nicholas C.Zakas'
                }
            ]
```

```
        },
        computed: {
            filterBooks: function () {
                return this.books.filter(function (book) {
                    return book.name.match(/JavaScript/);
                });
            }
        }
    })
</script>
```

上例是把书名中包含 JavaScript 关键词的数据过滤出来,计算属性 filterBooks 依赖 books,但是不会修改 books。实现排序也是类似的,比如在此基础上新加一个计算属性 sortedBooks,按照书名的长度由长到短进行排序:

```
computed: {
    sortedBooks: function () {
        return this.books.sort(function (a, b) {
            return a.name.length < b.name.length;
        });
    }
}
```

提示

在 Vue.js 2.x 中废弃了 1.x 中内置的 limitBy、filterBy 和 orderBy 过滤器,统一改用计算属性来实现。

5.4 方法与事件

5.4.1 基本用法

在第 2.2 节,我们已经引入了 Vue 事件处理的概念 v-on,在事件绑定上,类似原生 JavaScript 的 onclick 等写法,也是在 HTML 上进行监听的。例如,我们监听一个按钮的点击事件,设置一个计数器,每次点击都加 1:

```
<div id="app">
    点击次数:{{ counter }}
    <button @click="counter++">+ 1</button>
</div>
<script>
    new Vue({
        el: '#app',
        data: {
            counter: 0
```

```
            }
        })
    </script>
```

提示 上面的@click 等同于 v-on:click，是一个语法糖，如不特殊说明，后面都将使用语法糖写法，可以回顾第 2.3 节。

@click 的表达式可以直接使用 JavaScript 语句，也可以是一个在 Vue 实例中 methods 选项内的函数名，比如对上例进行扩展，再增加一个按钮，点击一次，计数器加 10：

```
<div id="app">
    点击次数：{{ counter }}
    <button @click="handleAdd()">+ 1</button>
    <button @click="handleAdd(10)">+ 10</button>
</div>
<script>
    var app = new Vue({
        el: '#app',
        data: {
            counter: 0
        },
        methods: {
            handleAdd: function (count) {
                count = count || 1;
                // this 指向当前 Vue 实例 app
                this.counter += count;
            }
        }
    })
</script>
```

在 methods 中定义了我们需要的方法供 @click 调用，需要注意的是，@click 调用的方法名后可以不跟括号“()”。此时，如果该方法有参数，默认会将原生事件对象 event 传入，可以尝试修改为@click="handleAdd"，然后在 handleAdd 内打印出 count 参数看看。在大部分业务场景中，如果方法不需要传入参数，为了简便可以不写括号。

这种在 HTML 元素上监听事件的设计看似将 DOM 与 JavaScript 紧耦合，违背分离的原理，实则刚好相反。因为通过 HTML 就可以知道调用的是哪个方法，将逻辑与 DOM 解耦，便于维护。最重要的是，当 ViewModel 销毁时，所有的事件处理器都会自动删除，无须自己清理。

Vue 提供了一个特殊变量$event，用于访问原生 DOM 事件，例如下面的实例可以阻止链接打开：

```
<div id="app">
    <a href="http://www.apple.com" @click="handleClick('禁止打开', $event)">
打开链接</a>
```

```
</div>
<script>
    var app = new Vue({
        el: '#app',
        methods: {
            handleClick: function (message, event) {
                event.preventDefault();
                window.alert(message);
            }
        }
    })
</script>
```

5.4.2 修饰符

在上例使用的 event.preventDefault()也可以用 Vue 事件的修饰符来实现，在@绑定的事件后加小圆点".",再跟一个后缀来使用修饰符。Vue 支持以下修饰符：

- .stop
- .prevent
- .capture
- .self
- .once

具体用法如下：

```
<!-- 阻止单击事件冒泡 -->
<a @click.stop="handle"></a>
<!-- 提交事件不再重载页面 -->
<form @submit.prevent="handle"></form>
<!--修饰符可以串联   -->
<a @click.stop.prevent="handle"></a>
<!-- 只有修饰符 -->
<form @submit.prevent></form>
<!--添加事件侦听器时使用事件捕获模式 -->
<div @click.capture="handle">...</div>
<!-- 只当事件在该元素本身（而不是子元素）触发时触发回调 -->
<div @click.self="handle">...</div>
<!-- 只触发一次，组件同样适用-->
<div @click.once="handle">...</div>
```

在表单元素上监听键盘事件时，还可以使用按键修饰符，比如按下具体某个键时才调用方法：

```
<!-- 只有在 keyCode 是 13 时调用 vm.submit() -->
<input @keyup.13="submit">
```

也可以自己配置具体按键：

```
Vue.config.keyCodes.f1 = 112;
// 全局定义后，就可以使用@keyup.f1
```

除了具体的某个 keyCode 外，Vue 还提供了一些快捷名称，以下是全部的别名：

- .enter
- .tab
- .delete（捕获"删除"和"退格"键）
- .esc
- .space
- .up
- .down
- .left
- .right

这些按键修饰符也可以组合使用，或和鼠标一起配合使用：

- .ctrl
- .alt
- .shift
- .meta（Mac 下是 Command 键，Windows 下是窗口键）

例如：

```
<!-- Shift + S -->
<input @keyup.shift.83="handleSave">
<!-- Ctrl + Click -->
<div @click.ctrl="doSomething">Do something</div>
```

以上就是事件指令 v-on 的基本内容，在第 7 章的组件中，我们还将介绍用 v-on 来绑定自定义事件。

5.5 实战：利用计算属性、指令等知识开发购物车

前 5 章内容基本涵盖了 Vue.js 最核心和常用的知识点，掌握这些内容已经可以上手开发一些小功能了。本节则以 Vue.js 的计算属性、内置指令、方法等内容为基础，完成一个在业务中具有代表性的小功能：购物车。

在开始写代码前，要对需求进行分析，这样有助于我们理清业务逻辑，尽可能还原设计与产品交互。

购物车需要展示一个已加入购物车的商品列表，包含商品名称、商品单价、购买数量和操作等信息，还需要实时显示购买的总价。其中购买数量可以增加或减少，每类商品还可以从购物车中

移除。最终实现的效果如图 5-6 所示。

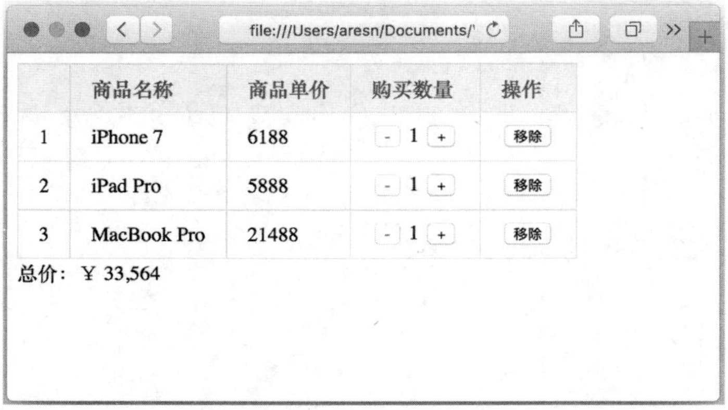

图 5-6　购物车效果图

在明确需求后，我们就可以开始编程了，因为业务代码较多，这次我们将 HTML、CSS、JavaScript 分离为 3 个文件，便于阅读和维护：

- index.html（引入资源及模板）
- index.js（Vue 实例及业务代码）
- style.css（样式）

先在 index.html 中引入 Vue.js 和相关资源，创建一个根元素来挂载 Vue 实例：

```html
<!DOCTYPE html>
<html>
<head>
    <meta charset="utf-8">
    <title>购物车示例</title>
    <link rel="stylesheet" type="text/css" href="style.css">
</head>
<body>
    <div id="app" v-cloak>

    </div>
    <script src="https://unpkg.com/vue/dist/vue.min.js"></script>
    <script src="index.js"></script>
</body>
</html>
```

注意，这里将 vue.min.js 和 index.js 文件写在<body>的最底部，如果写在<head>里，Vue 实例将无法创建，因为此时 DOM 还没有被解析完成，除非通过异步或在事件 DOMContentLoaded（IE 是 onreadystatechange）触发时再创建 Vue 实例，这有点像 jQuery 的$(document).ready()方法。

本例需要用到 Vue.js 的 computed、methods 等选项，在 index.js 中先初始化实例：

```
var app = new Vue({
```

```
    el: '#app',
    data: {

    },
    computed: {

    },
    methods: {

    }
});
```

我们需要的数据比较简单，只有一个列表，里面包含了商品名称、单价、购买数量。在实际业务中，这个列表应该是通过 Ajax 从服务端动态获取的，这里只做示例，所以直接写入在 data 选项内，另外每个商品还应该有一个全局唯一的 id。我们在 data 内写入列表 list：

```
data: {
    list: [
        {
            id: 1,
            name: 'iPhone 7',
            price: 6188,
            count: 1
        },
        {
            id: 2,
            name: 'iPad Pro',
            price: 5888,
            count: 1
        },
        {
            id: 3,
            name: 'MacBook Pro',
            price: 21488,
            count: 1
        }
    ]
}
```

数据构建好后，可以在 index.html 中展示列表了，毫无疑问，肯定会用到 v-for，不过在此之前，我们先做一些小的优化。因为每个商品都是可以从购物车移除的，所以当列表为空时，在页面中显示一个"购物车为空"的提示更为友好，我们可以通过判断数组 list 的长度来实现该功能：

```
<div id="app" v-cloak>
    <template v-if="list.length">
```

```
    </template>
    <div v-else>购物车为空</div>
</div>
```

<template>里的代码分两部分,一部分是商品列表信息,我们用表格 table 来展现;另一部分就是带有千位分隔符的商品总价(每隔三位数加进一个逗号)。这部分代码如下:

```
<template v-if="list.length">
    <table>
        <thead>
            <tr>
                <th></th>
                <th>商品名称</th>
                <th>商品单价</th>
                <th>购买数量</th>
                <th>操作</th>
            </tr>
        </thead>
        <tbody>

        </tbody>
    </table>
    <div>总价:¥ {{ totalPrice }}</div>
</template>
```

总价 totalPrice 是依赖于商品列表而动态变化的,所以我们用计算属性来实现,顺便将结果转换为带有"千位分隔符"的数字,在 index.js 的 computed 选项内写入:

```
computed: {
    totalPrice: function () {
        var total = 0;
        for (var i = 0; i < this.list.length; i++) {
            var item = this.list[i];
            total += item.price * item.count;
        }
        return total.toString().replace(/\B(?=(\d{3})+$)/g,',');
    }
}
```

这段代码难点在于千位分隔符的转换,读者可以查阅正则匹配的相关内容后尝试了解 replace() 的正则含义。

最后就剩下商品列表的渲染和相关的几个操作了。先在<tbody>内把数组 list 用 v-for 指令循环出来:

```
<tbody>
```

```
    <tr v-for="(item, index) in list">
        <td>{{ index + 1 }}</td>
        <td>{{ item.name }}</td>
        <td>{{ item.price }}</td>
        <td>
            <button
                @click="handleReduce(index)"
                :disabled="item.count === 1">-</button>
            {{ item.count }}
            <button @click="handleAdd(index)">+</button>
        </td>
        <td>
            <button @click="handleRemove(index)">移除</button>
        </td>
    </tr>
</tbody>
```

商品序号、名称、单价、数量都是直接使用插值来完成的，在第 4 列的两个按钮 <button> 用于增/减购买数量，分别绑定了两个方法 handleReduce 和 handleAdd，参数都是当前商品在数组 list 中的索引。很多时候，一个元素上会同时使用多个特性（尤其是在组件中使用 props 传递数据时），写在一行代码较长，不便阅读，所以建议特性过多时，将每个特性都单独写为一行，比如第一个 <button >中使用了 v-bind 和 v-on 两个指令（这里都用的语法糖写法）。每件商品购买数量最少是 1 件，所以当 count 为 1 时，不允许再继续减少，所以这里给<button>动态绑定了 disabled 特性来禁用按钮。

在 index.js 中继续完成剩余的 3 个方法：

```
methods: {
    handleReduce: function (index) {
        if (this.list[index].count === 1) return;
        this.list[index].count--;
    },
    handleAdd: function (index) {
        this.list[index].count++;
    },
    handleRemove: function (index) {
        this.list.splice(index, 1);
    }
}
```

这 3 个方法都是直接对数组 list 的操作，没有太复杂的逻辑。需要说明的是，虽然在 button 上已经绑定了 disabled 特性，但是在 handleReduce 方法内又判断了一遍，这是因为在某些时候，可能不一定会用 button 元素，也可能是 div、span 等，给它们增加 disabled 是没有任何作用的，所以安全起见，在业务逻辑中再判断一次，避免因修改 HTML 模板后出现 bug。

以下是购物车示例的完整代码：

index.html：

```html
<!DOCTYPE html>
<html>
<head>
    <meta charset="utf-8">
    <title>购物车示例</title>
    <link rel="stylesheet" type="text/css" href="style.css">
</head>
<body>
    <div id="app" v-cloak>
        <template v-if="list.length">
            <table>
                <thead>
                    <tr>
                        <th></th>
                        <th>商品名称</th>
                        <th>商品单价</th>
                        <th>购买数量</th>
                        <th>操作</th>
                    </tr>
                </thead>
                <tbody>
                    <tr v-for="(item, index) in list">
                        <td>{{ index + 1 }}</td>
                        <td>{{ item.name }}</td>
                        <td>{{ item.price }}</td>
                        <td>
                            <button
                                @click="handleReduce(index)"
                                :disabled="item.count === 1">-</button>
                            {{ item.count }}
                            <button @click="handleAdd(index)">+</button>
                        </td>
                        <td>
                            <button @click="handleRemove(index)">移除</button>
                        </td>
                    </tr>
                </tbody>
            </table>
            <div>总价：￥ {{ totalPrice }}</div>
        </template>
```

```html
        <div v-else>购物车为空</div>
    </div>
    <script src="https://unpkg.com/vue/dist/vue.min.js"></script>
    <script src="index.js"></script>
</body>
</html>
```

index.js：

```js
var app = new Vue({
    el: '#app',
    data: {
        list: [
            {
                id: 1,
                name: 'iPhone 7',
                price: 6188,
                count: 1
            },
            {
                id: 2,
                name: 'iPad Pro',
                price: 5888,
                count: 1
            },
            {
                id: 3,
                name: 'MacBook Pro',
                price: 21488,
                count: 1
            }
        ]
    },
    computed: {
        totalPrice: function () {
            var total = 0;
            for (var i = 0; i < this.list.length; i++) {
                var item = this.list[i];
                total += item.price * item.count;
            }

            return total.toString().replace(/\B(?=(\d{3})+$)/g, ',');
        }
    },
```

```
    methods: {
        handleReduce: function (index) {
            if (this.list[index].count === 1) return;
            this.list[index].count--;
        },
        handleAdd: function (index) {
            this.list[index].count++;
        },
        handleRemove: function (index) {
            this.list.splice(index, 1);
        }
    }
});
```

style.css:

```
[v-cloak] {
    display: none;
}
table{
    border: 1px solid #e9e9e9;
    border-collapse: collapse;
    border-spacing: 0;
    empty-cells: show;
}
th, td{
    padding: 8px 16px;
    border: 1px solid #e9e9e9;
    text-align: left;
}
th{
    background: #f7f7f7;
    color: #5c6b77;
    font-weight: 600;
    white-space: nowrap;
}
```

练习 1：在当前示例基础上扩展商品列表，新增一项是否选中该商品的功能，总价变为只计算选中商品的总价，同时提供一个全选的按钮。

练习 2：将商品列表 list 改为一个二维数组来实现商品的分类，比如可分为"电子产品""生活用品"和"果蔬"，同类商品聚合在一起。提示，你可能会用到两次 v-for。

第 6 章

表单与 v-model

表单类控件承载了一个网页数据的录入与交互，本章将介绍如何使用指令 v-model 完成表单的数据双向绑定。

6.1 基本用法

表单控件在实际业务较为常见，比如单选、多选、下拉选择、输入框等，用它们可以完成数据的录入、校验、提交等。Vue.js 提供了 v-model 指令，用于在表单类元素上双向绑定数据，例如在输入框上使用时，输入的内容会实时映射到绑定的数据上。例如下面的例子：

```
<div id="app">
    <input type="text" v-model="message" placeholder="输入...">
    <p>输入的内容是：{{ message }}</p>
</div>
<script>
    var app = new Vue({
        el: '#app',
        data: {
            message: ''
        }
    })
</script>
```

在输入框输入的同时，{{ message }} 也会实时将内容渲染在视图中，如图 6-1 所示。

> Hello World
>
> 输入的内容是：Hello World

图 6-1　v-model 指令对数据的双向绑定

对于文本域 textarea 也是同样的用法：

```
<div id="app">
    <textarea v-model="text" placeholder="输入..."></textarea>
    <p>输入的内容是：</p>
    <p style="white-space: pre">{{ text }}</p>
</div>
<script>
    var app = new Vue({
        el: '#app',
        data: {
            text: ''
        }
    })
</script>
```

提示

使用 v-model 后，表单控件显示的值只依赖所绑定的数据，不再关心初始化时的 value 属性，对于在<textarea></textarea> 之间插入的值，也不会生效。

使用 v-model 时，如果是用中文输入法输入中文，一般在没有选定词组前，也就是在拼音阶段，Vue 是不会更新数据的，当敲下汉字时才会触发更新。如果希望总是实时更新，可以用@input 来替代 v-model。事实上，v-model 也是一个特殊的语法糖，只不过它会在不同的表单上智能处理。例如下面的示例：

```
<div id="app">
    <input type="text" @input="handleInput" placeholder="输入...">
    <p>输入的内容是：{{ message }}</p>
</div>
<script>
    var app = new Vue({
        el: '#app',
        data: {
            message: ''
        },
        methods: {
            handleInput: function (e) {
```

```
                this.message = e.target.value;
            }
        }
    })
</script>
```

来看看更多的表单控件。

单选按钮：

单选按钮在单独使用时，不需要 v-model，直接使用 v-bind 绑定一个布尔类型的值，为真时选中，为否时不选，例如：

```
<div id="app">
    <input type="radio" :checked="picked">
    <label>单选按钮</label>
</div>
<script>
    var app = new Vue({
        el: '#app',
        data: {
            picked: true
        }
    })
</script>
```

如果是组合使用来实现互斥选择的效果，就需要 v-model 配合 value 来使用：

```
<div id="app">
    <input type="radio" v-model="picked" value="html" id="html">
    <label for="html">HTML</label>
    <br>
    <input type="radio" v-model="picked" value="js" id="js">
    <label for="js">JavaScript</label>
    <br>
    <input type="radio" v-model="picked" value="css" id="css">
    <label for="css">CSS</label>
    <br>
    <p>选择的项是：{{ picked }}</p>
</div>
<script>
    var app = new Vue({
        el: '#app',
        data: {
            picked: 'js'
        }
```

    })
</script>

数据 picked 的值与单选按钮的 value 值一致时，就会选中该项，所以当前状态下选中的是第二项 JavaScript，如图 6-2 所示。

图 6-2   单选按钮示例结果

复选框：

复选框也分单独使用和组合使用，不过用法稍与单选不同。复选框单独使用时，也是用 v-model 来绑定一个布尔值，例如：

```
<div id="app">
    <input type="checkbox" v-model="checked" id="checked">
    <label for="checked">选择状态：{{ checked }}</label>
</div>
<script>
    var app = new Vue({
        el: '#app',
        data: {
            checked: false
        }
    })
</script>
```

在勾选时，数据 checked 的值变为了 true，label 中渲染的内容也会更新。

组合使用时，也是 v-model 与 value 一起，多个勾选框都绑定到同一个数组类型的数据，value 的值在数组当中，就会选中这一项。这一过程也是双向的，在勾选时，value 的值也会自动 push 到这个数组中，示例代码如下：

```
<div id="app">
    <input type="checkbox" v-model="checked" value="html" id="html">
    <label for="html">HTML</label>
    <br>
    <input type="checkbox" v-model="checked" value="js" id="js">
    <label for="js">JavaScript</label>
```

```
        <br>
        <input type="checkbox" v-model="checked" value="css" id="css">
        <label for="css">CSS</label>
        <br>
        <p>选择的项是：{{ checked }}</p>
    </div>
    <script>
        var app = new Vue({
            el: '#app',
            data: {
                checked: ['html', 'css']
            }
        })
    </script>
```

当前状态下的结果如图 6-3 所示。

图 6-3　多选框组合使用的结果

选择列表：

选择列表就是下拉选择器，也是常见的表单控件，同样也分为单选和多选两种方式。先看一下单选的示例代码：

```
<div id="app">
    <select v-model="selected">
        <option>html</option>
        <option value="js">JavaScript</option>
        <option>css</option>
    </select>
    <p>选择的项是：{{ selected }}</p>
</div>
<script>
    var app = new Vue({
        el: '#app',
```

```
        data: {
            selected: 'html'
        }
    })
</script>
```

<option>是备选项，如果含有 value 属性，v-model 就会优先匹配 value 的值；如果没有，就会直接匹配 <option>的 text，比如选中第二项时，selected 的值是 js，而不是 JavaScript。

给<select>添加属性 multiple 就可以多选了，此时 v-model 绑定的是一个数组，与复选框用法类似，示例代码如下：

```
<div id="app">
    <select v-model="selected" multiple>
        <option>html</option>
        <option value="js">JavaScript</option>
        <option>css</option>
    </select>
    <p>选择的项是：{{ selected }}</p>
</div>
<script>
    var app = new Vue({
        el: '#app',
        data: {
            selected: ['html', 'js']
        }
    })
</script>
```

在业务中，<option>经常用 v-for 动态输出，value 和 text 也是用 v-bind 来动态输出的，例如：

```
<div id="app">
    <select v-model="selected">
        <option
            v-for="option in options"
            :value="option.value">{{ option.text }}</option>
    </select>
    <p>选择的项是：{{ selected }}</p>
</div>
<script>
    var app = new Vue({
        el: '#app',
        data: {
            selected: 'html',
            options: [
                {
```

```
                    text: 'HTML',
                    value: 'html'
                },
                {
                    text: 'JavaScript',
                    value: 'js'
                },
                {
                    text: 'CSS',
                    value: 'css'
                }
            ]
        }
    })
</script>
```

虽然用选择列表<select>控件可以很简单地完成下拉选择的需求，但是在实际业务中反而不常用，因为它的样式依赖平台和浏览器，无法统一，也不太美观，功能也受限，比如不支持搜索，所以常见的解决方案是用 div 模拟一个类似的控件。当阅读完第 7 章组件的内容后，可以尝试编写一个下拉选择器的通用组件。

6.2 绑定值

上一节介绍的单选按钮、复选框和选择列表在单独使用或单选的模式下，v-model 绑定的值是一个静态字符串或布尔值，但在业务中，有时需要绑定一个动态的数据，这时可以用 v-bind 来实现。

单选按钮：

```
<div id="app">
    <input type="radio" v-model="picked" :value="value">
    <label>单选按钮</label>
    <p>{{ picked }}</p>
    <p>{{ value }}</p>
</div>
<script>
    var app = new Vue({
        el: '#app',
        data: {
            picked: false,
            value: 123
        }
    })
</script>
```

在选中时，app.picked === app.value，值都是 123。

复选框：

```html
<div id="app">
    <input
        type="checkbox"
        v-model="toggle"
        :true-value="value1"
        :false-value="value2">
    <label>复选框</label>
    <p>{{ toggle }}</p>
    <p>{{ value1 }}</p>
    <p>{{ value2 }}</p>
</div>
<script>
    var app = new Vue({
        el: '#app',
        data: {
            toggle: false,
            value1: 'a',
            value2: 'b'
        }
    })
</script>
```

勾选时，app.toggle === app.value1；未勾选时，app.toggle === app.value2。

选择列表：

```html
<div id="app">
    <select v-model="selected">
        <option :value="{ number: 123 }">123</option>
    </select>
    {{ selected.number }}
</div>
<script>
    var app = new Vue({
        el: '#app',
        data: {
            selected: ''
        }
    })
</script>
```

当选中时，app.selected 是一个 Object，所以 app.selected.number === 123。

6.3 修饰符

与事件的修饰符类似，v-model 也有修饰符，用于控制数据同步的时机。

.lazy：

在输入框中，v-model 默认是在 input 事件中同步输入框的数据（除了提示中介绍的中文输入法情况外），使用修饰符 .lazy 会转变为在 change 事件中同步，示例代码如下：

```
<div id="app">
    <input type="text" v-model.lazy="message">
    <p>{{ message }}</p>
</div>
<script>
    var app = new Vue({
        el: '#app',
        data: {
            message: ''
        }
    })
</script>
```

这时，message 并不是实时改变的，而是在失焦或按回车时才更新。

.number：

使用修饰符.number 可以将输入转换为 Number 类型，否则虽然你输入的是数字，但它的类型其实是 String，比如在数字输入框时会比较有用，示例代码如下：

```
<div id="app">
    <input type="number" v-model.number="message">
    <p>{{ typeof message }}</p>
</div>
<script>
    var app = new Vue({
        el: '#app',
        data: {
            message: 123
        }
    })
</script>
```

.trim：

修饰符 .trim 可以自动过滤输入的首尾空格，示例代码如下：

```
<div id="app">
    <input type="text" v-model.trim="message">
    <p>{{ message }}</p>
</div>
<script>
    var app = new Vue({
        el: '#app',
        data: {
            message: ''
        }
    })
</script>
```

从 Vue.js 2.x 开始，v-model 还可以用于自定义组件，满足定制化的需求，在第 7 章会详细介绍。

第 7 章

组 件 详 解

组件（Component）是 Vue.js 最核心的功能，也是整个框架设计最精彩的地方，当然也是最难掌握的。本章将带领你由浅入深地学习组件的全部内容，并通过几个实战项目熟练使用 Vue 组件。

7.1 组件与复用

7.1.1 为什么使用组件

在正式介绍组件前，我们先来看一个简单的场景，如图 7-1 所示。

图 7-1 常见的聊天界面

图 7-1 中是一个很常见的聊天界面，有一些标准的控件，比如右上角的关闭按钮、输入框、发送按钮等。你可能要问了，这有什么难的，不就是几个 div、input 吗？好，那现在需求升级了，这几个控件还有别的地方要用到。没问题，复制粘贴呗。那如果输入框要带数据验证，按钮的图标支持自定义呢？这样用 JavaScript 封装后一起复制吧。那等到项目快完结时，产品经理说，所有使用输入框的地方，都要改成支持回车键提交。好吧，给我一天的时间，我一个一个加上去。

上面的需求虽然有点变态，但却是业务中很常见的，那就是一些控件、JavaScript 能力的复用。没错，Vue.js 的组件就是提高重用性的，让代码可复用，当学习完组件后，上面的问题就可以分分钟搞定了，再也不用害怕产品经理的奇葩需求。

我们先看一下图 7-1 中的示例用组件来编写是怎样的，示例代码如下：

```html
<Card style="width: 350px;">
    <p slot="title">与 xxx 聊天中</p>
    <a href="#" slot="extra">
        <Icon type="android-close" size="18"></Icon>
    </a>
    <div style="height: 100px;">

    </div>
    <div>
        <Row :gutter="16">
            <i-col span="17">
                <i-input
                    v-model="value"
                    placeholder="请输入..."></i-input>
            </i-col>
            <i-col span="4">
                <i-button
                    type="primary"
                    icon="paper-airplane">发送</i-button>
            </i-col>
        </Row>
    </div>
</Card>
```

是不是很奇怪，有很多我们从来都没有见过的标签，比如<Card>、<Row>、<i-col>、<i-input>和<i-button>等，而且整段代码除了内联的几个样式外，一句 CSS 代码也没有，但最终实现的 UI 就是图 7-1 的效果。

这些没见过的自定义标签就是组件，每个标签代表一个组件，在任何使用 Vue 的地方都可以直接使用。接下来，我们就来看看组件的具体用法。

7.1.2 组件用法

回顾一下我们创建 Vue 实例的方法：

```
var app = new Vue({
    el: '#app'
})
```

组件与之类似，需要注册后才可以使用。注册有全局注册和局部注册两种方式。全局注册后，任何 Vue 实例都可以使用。全局注册示例代码如下：

```
Vue.component('my-component', {
    //选项
})
```

my-component 就是注册的组件自定义标签名称，推荐使用小写加减号分割的形式命名。

要在父实例中使用这个组件，必须要在实例创建前注册，之后就可以用<my-component></my-component>的形式来使用组件了，示例代码如下：

```
<div id="app">
    <my-component></my-component>
</div>
<script>
    Vue.component('my-component', {
        //选项
    });

    var app = new Vue({
        el: '#app'
    })
</script>
```

此时打开页面还是空白的，因为我们注册的组件没有任何内容，在组件选项中添加 template 就可以显示组件内容了，示例代码如下：

```
Vue.component('my-component', {
    template: '<div>这里是组件的内容</div>'
});
```

渲染后的结果是：

```
<div id="app">
    <div>这里是组件的内容</div>
</div>
```

template 的 DOM 结构必须被一个元素包含，如果直接写成"这里是组件的内容"，不带"<div></div>"是无法渲染的。

在 Vue 实例中，使用 components 选项可以局部注册组件，注册后的组件只有在该实例作用域下有效。组件中也可以使用 components 选项来注册组件，使组件可以嵌套。示例代码如下：

```
<div id="app">
    <my-component></my-component>
</div>
<script>
    var Child = {
        template: '<div>局部注册组件的内容</div>'
    }

    var app = new Vue({
        el: '#app',
        components: {
            'my-component': Child
        }
    })
</script>
```

Vue 组件的模板在某些情况下会受到 HTML 的限制，比如<table>内规定只允许是<tr>、<td>、<th>等这些表格元素，所以在<table>内直接使用组件是无效的。这种情况下，可以使用特殊的 is 属性来挂载组件，示例代码如下：

```
<div id="app">
    <table>
        <tbody is="my-component"></tbody>
    </table>
</div>
<script>
    Vue.component('my-component', {
        template: '<div>这里是组件的内容</div>'
    });
    var app = new Vue({
        el: '#app'
    })
</script>
```

tbody 在渲染时，会被替换为组件的内容。常见的限制元素还有<ul>、<ol>、<select>。

如果使用的是字符串模板，是不受限制的，比如后面章节介绍的.vue 单文件用法等。

除了 template 选项外，组件中还可以像 Vue 实例那样使用其他的选项，比如 data、computed、methods 等。但是在使用 data 时，和实例稍有区别，data 必须是函数，然后将数据 return 出去，例如：

```
<div id="app">
    <my-component></my-component>
```

```
    </div>
    <script>
        Vue.component('my-component', {
            template: '<div>{{ message }}</div>',
            data: function () {
                return {
                    message: '组件内容'
                }
            }
        });

        var app = new Vue({
            el: '#app'
        })
    </script>
```

JavaScript 对象是引用关系，所以如果 return 出的对象引用了外部的一个对象，那这个对象就是共享的，任何一方修改都会同步。比如下面的示例：

```
<div id="app">
    <my-component></my-component>
    <my-component></my-component>
    <my-component></my-component>
</div>
<script>
    var data = {
        counter: 0
    };

    Vue.component('my-component', {
        template: '<button @click="counter++">{{ counter }}</button>',
        data: function () {
            return data;
        }
    });

    var app = new Vue({
        el: '#app'
    })
</script>
```

组件使用了 3 次，但是点击任意一个<button>，3 个的数字都会加 1，那是因为组件的 data 引用的是外部的对象，这肯定不是我们期望的效果，所以给组件返回一个新的 data 对象来独立，示例代码如下：

```
<div id="app">
    <my-component></my-component>
    <my-component></my-component>
    <my-component></my-component>
</div>
<script>
    Vue.component('my-component', {
        template: '<button @click="counter++">{{ counter }}</button>',
        data: function () {
            return {
                counter: 0
            };
        }
    });

    var app = new Vue({
        el: '#app'
    })
</script>
```

这样，点击 3 个按钮就互不影响了，完全达到复用的目的。

7.2 使用 props 传递数据

7.2.1 基本用法

组件不仅仅是要把模板的内容进行复用，更重要的是组件间要进行通信。通常父组件的模板中包含子组件，父组件要正向地向子组件传递数据或参数，子组件接收到后根据参数的不同来渲染不同的内容或执行操作。这个正向传递数据的过程就是通过 props 来实现的。

在组件中，使用选项 props 来声明需要从父级接收的数据，props 的值可以是两种，一种是字符串数组，一种是对象，本小节先介绍数组的用法。比如我们构造一个数组，接收一个来自父级的数据 message，并把它在组件模板中渲染，示例代码如下：

```
<div id="app">
    <my-component message="来自父组件的数据"></my-component>
</div>
<script>
    Vue.component('my-component', {
        props: ['message'],
        template: '<div>{{ message }}</div>'
    });
```

```
    var app = new Vue({
        el: '#app'
    })
</script>
```

渲染后的结果为：

```
<div id="app">
    <div>来自父组件的数据</div>
</div>
```

props 中声明的数据与组件 data 函数 return 的数据主要区别就是 props 的来自父级，而 data 中的是组件自己的数据，作用域是组件本身，这两种数据都可以在模板 template 及计算属性 computed 和方法 methods 中使用。上例的数据 message 就是通过 props 从父级传递过来的，在组件的自定义标签上直接写该 props 的名称，如果要传递多个数据，在 props 数组中添加项即可。

由于 HTML 特性不区分大小写，当使用 DOM 模板时，驼峰命名（camelCase）的 props 名称要转为短横分隔命名（kebab-case），例如：

```
<div id="app">
    <my-component warning-text="提示信息"></my-component>
</div>
<script>
    Vue.component('my-component', {
        props: ['warningText'],
        template: '<div>{{ warningText }}</div>'
    });

    var app = new Vue({
        el: '#app'
    })
</script>
```

提示

如果使用的是字符串模板，仍然可以忽略这些限制。

有时候，传递的数据并不是直接写死的，而是来自父级的动态数据，这时可以使用指令 v-bind 来动态绑定 props 的值，当父组件的数据变化时，也会传递给子组件。示例代码如下：

```
<div id="app">
    <input type="text" v-model="parentMessage">
    <my-component :message="parentMessage"></my-component>
</div>
<script>
    Vue.component('my-component', {
        props: ['message'],
```

```
        template: '<div>{{ message }}</div>'
    });

    var app = new Vue({
        el: '#app',
        data: {
            parentMessage: ''
        }
    })
</script>
```

这里用 v-model 绑定了父级的数据 parentMessage，当通过输入框任意输入时，子组件接收到的 props "message" 也会实时响应，并更新组件模板。

注意，如果你要直接传递数字、布尔值、数组、对象，而且不使用 v-bind，传递的仅仅是字符串，尝试下面的示例来对比：

```
<div id="app">
    <my-component message="[1,2,3]"></my-component>
    <my-component :message="[1,2,3]"></my-component>
</div>
<script>
    Vue.component('my-component', {
        props: ['message'],
        template: '<div>{{ message.length }}</div>'
    });

    var app = new Vue({
        el: '#app'
    })
</script>
```

同一个组件使用了两次，区别仅仅是第二个使用的是 v-bind。渲染后的结果，第一个是 7，第二个才是数组的长度 3。

7.2.2 单向数据流

Vue 2.x 与 Vue 1.x 比较大的一个改变就是，Vue 2.x 通过 props 传递数据是单向的了，也就是父组件数据变化时会传递给子组件，但是反过来不行。而在 Vue 1.x 里提供了 .sync 修饰符来支持双向绑定。之所以这样设计，是尽可能将父子组件解耦，避免子组件无意中修改了父组件的状态。

业务中会经常遇到两种需要改变 prop 的情况，一种是父组件传递初始值进来，子组件将它作为初始值保存起来，在自己的作用域下可以随意使用和修改。这种情况可以在组件 data 内再声明一个数据，引用父组件的 prop，示例代码如下：

```
<div id="app">
    <my-component :init-count="1"></my-component>
</div>
<script>
    Vue.component('my-component', {
        props: ['initCount'],
        template: '<div>{{ count }}</div>',
        data: function () {
            return {
                count: this.initCount
            }
        }
    });

    var app = new Vue({
        el: '#app'
    })
</script>
```

组件中声明了数据 count,它在组件初始化时会获取来自父组件的 initCount,之后就与之无关了,只用维护 count,这样就可以避免直接操作 initCount。

另一种情况就是 prop 作为需要被转变的原始值传入。这种情况用计算属性就可以了,示例代码如下:

```
<div id="app">
    <my-component :width="100"></my-component>
</div>
<script>
    Vue.component('my-component', {
        props: ['width'],
        template: '<div :style="style">组件内容</div>',
        computed: {
            style: function () {
                return {
                    width: this.width + 'px'
                }
            }
        }
    });

    var app = new Vue({
        el: '#app'
    })
</script>
```

因为用 CSS 传递宽度要带单位（px），但是每次都写太麻烦，而且数值计算一般是不带单位的，所以统一在组件内使用计算属性就可以了。

 注意，在 JavaScript 中对象和数组是引用类型，指向同一个内存空间，所以 props 是对象和数组时，在子组件内改变是会影响父组件的。

7.2.3 数据验证

我们上面所介绍的 props 选项的值都是一个数组，一开始也介绍过，除了数组外，还可以是对象，当 prop 需要验证时，就需要对象写法。

一般当你的组件需要提供给别人使用时，推荐都进行数据验证，比如某个数据必须是数字类型，如果传入字符串，就会在控制台弹出警告。

以下是几个 prop 的示例：

```
Vue.component('my-component', {
    props: {
        //必须是数字类型
        propA: Number,
        //必须是字符串或数字类型
        propB: [String, Number],
        // 布尔值，如果没有定义，默认值就是 true
        propC: {
            type: Boolean,
            default: true
        },
        //数字，而且是必传
        propD: {
            type: Number,
            required: true
        },
        // 如果是数组或对象，默认值必须是一个函数来返回
        propE: {
            type: Array,
            default: function () {
                return [];
            }
        },
        // 自定义一个验证函数
        propF: {
            validator: function (value) {
                return value > 10;
            }
        }
    }
});
```

验证的 type 类型可以是：

- String
- Number
- Boolean
- Object
- Array
- Function

type 也可以是一个自定义构造器，使用 instanceof 检测。

当 prop 验证失败时，在开发版本下会在控制台抛出一条警告。

7.3 组件通信

我们已经知道，从父组件向子组件通信，通过 props 传递数据就可以了，但 Vue 组件通信的场景不止有这一种，归纳起来，组件之间通信可以用图 7-2 表示。

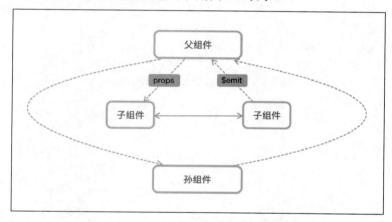

图 7-2 组件通信示例

组件关系可分为父子组件通信、兄弟组件通信、跨级组件通信。本节将介绍各种组件之间通信的方法。

7.3.1 自定义事件

当子组件需要向父组件传递数据时，就要用到自定义事件。我们在介绍指令 v-on 时有提到，v-on 除了监听 DOM 事件外，还可以用于组件之间的自定义事件。

如果你了解过 JavaScript 的设计模式——观察者模式，一定知道 dispatchEvent 和 addEventListener 这两个方法。Vue 组件也有与之类似的一套模式，子组件用$emit()来触发事件，父组件用$on()来监听子组件的事件。

父组件也可以直接在子组件的自定义标签上使用 v-on 来监听子组件触发的自定义事件，示例代码如下：

```html
<div id="app">
    <p>总数：{{ total }}</p>
    <my-component
        @increase="handleGetTotal"
        @reduce="handleGetTotal"></my-component>
</div>
<script>
    Vue.component('my-component', {
        template: '\
        <div>\
            <button @click="handleIncrease">+1</button>\
            <button @click="handleReduce">-1</button>\
        </div>',
        data: function () {
            return {
                counter: 0
            }
        },
        methods: {
            handleIncrease: function () {
                this.counter++;
                this.$emit('increase', this.counter);
            },
            handleReduce: function () {
                this.counter--;
                this.$emit('reduce', this.counter);
            }
        }
    });

    var app = new Vue({
        el: '#app',
        data: {
            total: 0
        },
        methods: {
            handleGetTotal: function (total) {
                this.total = total;
            }
        }
    })
</script>
```

上面示例中，子组件有两个按钮，分别实现加 1 和减 1 的效果，在改变组件的 data "counter"后，通过$emit()再把它传递给父组件，父组件用 v-on:increase 和 v-on:reduce（示例使用的是语法糖）。$emit()方法的第一个参数是自定义事件的名称，例如示例的 increase 和 reduce 后面的参数都是要传递的数据，可以不填或填写多个。

除了用 v-on 在组件上监听自定义事件外，也可以监听 DOM 事件，这时可以用.native 修饰符表示监听的是一个原生事件，监听的是该组件的根元素，示例代码如下：

```
<my-component v-on:click.native="handleClick"></my-component>
```

7.3.2 使用 v-model

Vue 2.x 可以在自定义组件上使用 v-model 指令，我们先来看一个示例：

```
<div id="app">
    <p>总数：{{ total }}</p>
    <my-component v-model="total"></my-component>
</div>
<script>
    Vue.component('my-component', {
        template: '<button @click="handleClick">+1</button>',
        data: function () {
            return {
                counter: 0
            }
        },
        methods: {
            handleClick: function () {
                this.counter++;
                this.$emit('input', this.counter);
            }
        }
    });

    var app = new Vue({
        el: '#app',
        data: {
            total: 0
        }
    })
</script>
```

仍然是点击按钮加 1 的效果，不过这次组件$emit()的事件名是特殊的 input，在使用组件的父级，并没有在<my-component>上使用@input= "handler"，而是直接用了 v-model 绑定的一个数据 total。这也可以称作是一个语法糖，因为上面的示例可以间接地用自定义事件来实现：

```
<div id="app">
    <p>总数：{{ total }}</p>
    <my-component @input="handleGetTotal"></my-component>
</div>
<script>
    // ...省略组件代码
    var app = new Vue({
        el: '#app',
        data: {
            total: 0
        },
        methods: {
            handleGetTotal: function (total) {
                this.total = total;
            }
        }
    })
</script>
```

v-model 还可以用来创建自定义的表单输入组件，进行数据双向绑定，例如：

```
<div id="app">
    <p>总数：{{ total }}</p>
    <my-component v-model="total"></my-component>
    <button @click="handleReduce">-1</button>
</div>
<script>
    Vue.component('my-component', {
        props: ['value'],
        template: '<input :value="value" @input="updateValue">',
        methods: {
            updateValue: function (event) {
                this.$emit('input', event.target.value);
            }
        }
    });
    var app = new Vue({
        el: '#app',
        data: {
            total: 0
        },
        methods: {
```

```
                handleReduce: function () {
                    this.total--;
                }
            }
        })
    </script>
```

实现这样一个具有双向绑定的 v-model 组件要满足下面两个要求：

- 接收一个 value 属性。
- 在有新的 value 时触发 input 事件。

7.3.3 非父子组件通信

在实际业务中，除了父子组件通信外，还有很多非父子组件通信的场景，非父子组件一般有两种，兄弟组件和跨多级组件。为了更加彻底地了解 Vue.js 2.x 中的通信方法，我们先来看一下在 Vue.js 1.x 中是如何实现的，这样便于我们了解 Vue.js 的设计思想。

在 Vue.js 1.x 中，除了$emit()方法外，还提供了$dispatch()和$broadcast() 这两个方法。$dispatch()用于向上级派发事件，只要是它的父级（一级或多级以上），都可以在 Vue 实例的 events 选项内接收，示例代码如下：

```
<!-- 注意：该示例需使用 Vue.js 1.x 的版本-->
<div id="app">
    {{ message }}
    <my-component></my-component>
</div>
<script>
    Vue.component('my-component', {
        template: '<button @click="handleDispatch">派发事件</button>',
        methods: {
            handleDispatch: function () {
                this.$dispatch('on-message', '来自内部组件的数据');
            }
        }
    });
    var app = new Vue({
        el: '#app',
        data: {
            message: ''
        },
        events: {
            'on-message': function (msg) {
                this.message = msg;
            }
```

```
        }
    })
</script>
```

同理，$broadcast()是由上级向下级广播事件的，用法完全一致，只是方向相反。

这两种方法一旦发出事件后，任何组件都是可以接收到的，就近原则，而且会在第一次接收到后停止冒泡，除非返回 true。

这两个方法虽然看起来很好用，但是在 Vue.js 2.x 中都废弃了，因为基于组件树结构的事件流方式让人难以理解，并且在组件结构扩展的过程中会变得越来越脆弱，并且不能解决兄弟组件通信的问题。

在 Vue.js 2.x 中，推荐使用一个空的 Vue 实例作为中央事件总线（bus），也就是一个中介。为了更形象地了解它，我们举一个生活中的例子。

比如你需要租房子，你可能会找房产中介来登记你的需求，然后中介把你的信息发给满足要求的出租者，出租者再把报价和看房时间告诉中介，由中介再转达给你，整个过程中，买家和卖家并没有任何交流，都是通过中间人来传话的。

或者你最近可能要换房了，你会找房产中介登记你的信息，订阅与你找房需求相关的资讯，一旦有符合你的房子出现时，中介会通知你，并传达你房子的具体信息。

这两个例子中，你和出租者担任的就是两个跨级的组件，而房产中介就是这个中央事件总线（bus）。比如下面的示例代码：

```
<div id="app">
    {{ message }}
    <component-a></component-a>
</div>
<script>
    var bus = new Vue();

    Vue.component('component-a', {
        template: '<button @click="handleEvent">传递事件</button>',
        methods: {
            handleEvent: function () {
                bus.$emit('on-message', '来自组件 component-a 的内容');
            }
        }
    });

    var app = new Vue({
        el: '#app',
        data: {
            message: ''
        },
        mounted: function () {
```

```
            var _this = this;
            //在实例初始化时，监听来自 bus 实例的事件
            bus.$on('on-message', function (msg) {
                _this.message = msg;
            });
        }
    })
</script>
```

首先创建了一个名为 bus 的空 Vue 实例，里面没有任何内容；然后全局定义了组件 component-a；最后创建 Vue 实例 app，在 app 初始化时，也就是在生命周期 mounted 钩子函数里监听了来自 bus 的事件 on-message，而在组件 component-a 中，点击按钮会通过 bus 把事件 on-message 发出去，此时 app 就会接收到来自 bus 的事件，进而在回调里完成自己的业务逻辑。

这种方法巧妙而轻量地实现了任何组件间的通信，包括父子、兄弟、跨级，而且 Vue 1.x 和 Vue 2.x 都适用。如果深入使用，可以扩展 bus 实例，给它添加 data、methods、computed 等选项，这些都是可以公用的，在业务中，尤其是协同开发时非常有用，因为经常需要共享一些通用的信息，比如用户登录的昵称、性别、邮箱等，还有用户的授权 token 等。只需在初始化时让 bus 获取一次，任何时间、任何组件就可以从中直接使用了，在单页面富应用（SPA）中会很实用，我们会在进阶篇里逐步介绍这些内容。

当你的项目比较大，有更多的小伙伴参与开发时，也可以选择更好的状态管理解决方案 vuex，在进阶篇里会详细介绍关于它的用法。

除了中央事件总线 bus 外，还有两种方法可以实现组件间通信：父链和子组件索引。

父链

在子组件中，使用 this.$parent 可以直接访问该组件的父实例或组件，父组件也可以通过 this.$children 访问它所有的子组件，而且可以递归向上或向下无限访问，直到根实例或最内层的组件。示例代码如下：

```
<div id="app">
    {{ message }}
    <component-a></component-a>
</div>
<script>
    Vue.component('component-a', {
        template: '<button @click="handleEvent">通过父链直接修改数据</button>',
        methods: {
            handleEvent: function () {
                // 访问到父链后，可以做任何操作，比如直接修改数据
                this.$parent.message = '来自组件 component-a 的内容';
            }
        }
    });
```

```
    var app = new Vue({
        el: '#app',
        data: {
            message: ''
        }
    })
</script>
```

尽管 Vue 允许这样操作，但在业务中，子组件应该尽可能地避免依赖父组件的数据，更不应该去主动修改它的数据，因为这样使得父子组件紧耦合，只看父组件，很难理解父组件的状态，因为它可能被任意组件修改，理想情况下，只有组件自己能修改它的状态。父子组件最好还是通过 props 和 $emit 来通信。

子组件索引

当子组件较多时，通过 this.$children 来一一遍历出我们需要的一个组件实例是比较困难的，尤其是组件动态渲染时，它们的序列是不固定的。Vue 提供了子组件索引的方法，用特殊的属性 ref 来为子组件指定一个索引名称，示例代码如下：

```
<div id="app">
    <button @click="handleRef">通过 ref 获取子组件实例</button>
    <component-a ref="comA"></component-a>
</div>
<script>
    Vue.component('component-a', {
        template: '<div>子组件</div>',
        data: function () {
            return {
                message: '子组件内容'
            }
        }
    });

    var app = new Vue({
        el: '#app',
        methods: {
            handleRef: function () {
                // 通过$refs 来访问指定的实例
                var msg = this.$refs.comA.message;
                console.log(msg);
            }
        }
    })
</script>
```

在父组件模板中，子组件标签上使用 ref 指定一个名称，并在父组件内通过 this.$refs 来访问指定名称的子组件。

提示
$refs 只在组件渲染完成后才填充，并且它是非响应式的。它仅仅作为一个直接访问子组件的应急方案，应当避免在模板或计算属性中使用$refs。

与 Vue 1.x 不同的是，Vue 2.x 将 v-el 和 v-ref 合并为了 ref，Vue 会自动去判断是普通标签还是组件。可以尝试补全下面的代码，分别打印出两个 ref 看看都是什么：

```
<div id="app">
    <p ref="p">内容</p>
    <child-component ref="child"></child-component>
</div>
```

7.4　使用 slot 分发内容

7.4.1　什么是 slot

我们先看一个比较常规的网站布局，如图 7-3 所示。

图 7-3　网站布局

这个网站由一级导航、二级导航、左侧列表、正文以及底部版权信息 5 个模块组成，如果要将它们都组件化，这个结构可能会是：

```
<app>
    <menu-main></menu-main>
    <menu-sub></menu-sub>
```

```
<div class="container">
    <menu-left></menu-left>
    <container></container>
</div>
<app-footer></app-footer>
</app>
```

当需要让组件组合使用，混合父组件的内容与子组件的模板时，就会用到 slot，这个过程叫作内容分发（transclusion）。以<app>为例，它有两个特点：

- <app>组件不知道它的挂载点会有什么内容。挂载点的内容是由<app>的父组件决定的。
- <app>组件很可能有它自己的模板。

props 传递数据、events 触发事件和 slot 内容分发就构成了 Vue 组件的 3 个 API 来源，再复杂的组件也是由这 3 部分构成的。

7.4.2 作用域

正式介绍 slot 前，需要先知道一个概念：编译的作用域。比如父组件中有如下模板：

```
<child-component>
    {{ message }}
</child-component>
```

这里的 message 就是一个 slot，但是它绑定的是父组件的数据，而不是组件<child-component>的数据。

父组件模板的内容是在父组件作用域内编译，子组件模板的内容是在子组件作用域内编译。例如下面的代码示例：

```
<div id="app">
    <child-component v-show="showChild"></child-component>
</div>
<script>
    Vue.component('child-component', {
        template: '<div>子组件</div>'
    });

    var app = new Vue({
        el: '#app',
        data: {
            showChild: true
        }
    })
</script>
```

这里的状态 showChild 绑定的是父组件的数据，如果想在子组件上绑定，那应该是：

```
<div id="app">
    <child-component></child-component>
</div>
<script>
    Vue.component('child-component', {
        template: '<div v-show="showChild">子组件</div>',
        data: function () {
            return {
                showChild: true
            }
        }
    });

    var app = new Vue({
        el: '#app'
    })
</script>
```

因此,slot 分发的内容,作用域是在父组件上的。

7.4.3　slot 用法

单个 Slot

在子组件内使用特殊的<slot>元素就可以为这个子组件开启一个 slot(插槽),在父组件模板里,插入在子组件标签内的所有内容将替代子组件的<slot> 标签及它的内容。示例代码如下:

```
<div id="app">
    <child-component>
        <p>分发的内容</p>
        <p>更多分发的内容</p>
    </child-component>
</div>
<script>
    Vue.component('child-component', {
        template: '\
        <div>\
            <slot>\
                <p>如果父组件没有插入内容,我将作为默认出现</p>\
            </slot>\
        </div>'
    });

    var app = new Vue({
        el: '#app'
```

```
    })
</script>
```

子组件 child-component 的模板内定义了一个<slot>元素,并且用一个<p>作为默认的内容,在父组件没有使用 slot 时,会渲染这段默认的文本;如果写入了 slot,那就会替换整个<slot>。所以上例渲染后的结果为:

```
<div id="app">
    <div>
        <p>分发的内容</p>
        <p>更多分发的内容</p>
    </div>
</div>
```

 注意,子组件<slot>内的备用内容,它的作用域是子组件本身。

具名 Slot

给<slot>元素指定一个 name 后可以分发多个内容,具名 Slot 可以与单个 Slot 共存,例如下面的示例:

```
<div id="app">
    <child-component>
        <h2 slot="header">标题</h2>
        <p>正文内容</p>
        <p>更多的正文内容</p>
        <div slot="footer">底部信息</div>
    </child-component>
</div>
<script>
    Vue.component('child-component', {
        template: '\
        <div class="container">\
            <div class="header">\
                <slot name="header"></slot>\
            </div>\
            <div class="main">\
                <slot></slot>\
            </div>\
            <div class="footer">\
                <slot name="footer"></slot>\
            </div>\
        </div>'
    });
```

```
    var app = new Vue({
        el: '#app'
    })
</script>
```

子组件内声明了 3 个<slot>元素，其中在<div class="main">内的<slot>没有使用 name 特性，它将作为默认 slot 出现，父组件没有使用 slot 特性的元素与内容都将出现在这里。

如果没有指定默认的匿名 slot，父组件内多余的内容片段都将被抛弃。

上例最终渲染后的结果为：

```
<div id="app">
    <div class="container">
        <div class="header">
            <h2>标题</h2>
        </div>
        <div class="main">
            <p>正文内容</p>
            <p>更多的正文内容</p>
        </div>
        <div class="footer">
            <div>底部信息</div>
        </div>
    </div>
</div>
```

在组合使用组件时，内容分发 API 至关重要。

7.4.4 作用域插槽

作用域插槽是一种特殊的 slot，使用一个可以复用的模板替换已渲染元素。概念比较难理解，我们先看一个简单的示例来了解它的基本用法。示例代码如下：

```
<div id="app">
    <child-component>
        <template scope="props">
            <p>来自父组件的内容</p>
            <p>{{ props.msg }}</p>
        </template>
    </child-component>
</div>
<script>
    Vue.component('child-component', {
        template: '\
        <div class="container">\
            <slot msg="来自子组件的内容"></slot>\
```

```
        </div>'
    });

    var app = new Vue({
        el: '#app'
    })
</script>
```

观察子组件的模板，在<slot>元素上有一个类似 props 传递数据给组件的写法 msg="xxx"，将数据传到了插槽。父组件中使用了<template>元素，而且拥有一个 scope="props"的特性，这里的 props 只是一个临时变量，就像 v-for="item in items"里面的 item 一样。template 内可以通过临时变量 props 访问来自子组件插槽的数据 msg。

将上面的示例渲染后的最终结果为：

```
<div id="app">
    <div class="container">
        <p>来组父组件的内容</p>
        <p>来自子组件的内容</p>
    </div>
</div>
```

作用域插槽更具代表性的用例是列表组件，允许组件自定义应该如何渲染列表每一项。示例代码如下：

```
<div id="app">
    <my-list :books="books">
        <!-- 作用域插槽也可以是具名的 Slot -->
        <template slot="book" scope="props">
            <li>{{ props.bookName }}</li>
        </template>
    </my-list>
</div>
<script>
    Vue.component('my-list', {
        props: {
            books: {
                type: Array,
                default: function () {
                    return [];
                }
            }
        },
        template: '\
        <ul>\
            <slot name="book"\
```

```
                    v-for="book in books"\
                    :book-name="book.name">\
                <!-- 这里也可以写默认 slot 内容 -->\
                </slot>\
            </ul>'
    });

    var app = new Vue({
        el: '#app',
        data: {
            books: [
                { name: '《Vue.js 实战》' },
                { name: '《JavaScript 语言精粹》' },
                { name: '《JavaScript 高级程序设计》' }
            ]
        }
    })
</script>
```

子组件 my-list 接收一个来自父级的 prop 数组 books，并且将它在 name 为 book 的 slot 上使用 v-for 指令循环，同时暴露一个变量 bookName。

如果你仔细揣摩上面的用法，你可能会产生这样的疑问：我直接在父组件用 v-for 不就好了吗，为什么还要绕一步，在子组件里面循环呢？的确，如果只是针对上面的示例，这样写是多此一举的。此例的用意主要是介绍作用域插槽的用法，并没有加入使用场景，而作用域插槽的使用场景就是既可以复用子组件的 slot，又可以使 slot 内容不一致。如果上例还在其他组件内使用，<li>的内容渲染权是由使用者掌握的，而数据却可以通过临时变量（比如 props）从子组件内获取。

7.4.5 访问 slot

在 Vue.js 1.x 中，想要获取某个 slot 是比较麻烦的，需要用 v-el 间接获取。而 Vue.js 2.x 提供了用来访问被 slot 分发的内容的方法 $slots，请看下面的示例：

```
<div id="app">
    <child-component>
        <h2 slot="header">标题</h2>
        <p>正文内容</p>
        <p>更多的正文内容</p>
        <div slot="footer">底部信息</div>
    </child-component>
</div>
<script>
    Vue.component('child-component', {
        template: '\
            <div class="container">\
```

```
            <div class="header">\
                <slot name="header"></slot>\
            </div>\
            <div class="main">\
                <slot></slot>\
            </div>\
            <div class="footer">\
                <slot name="footer"></slot>\
            </div>',
        mounted: function () {
            var header = this.$slots.header;
            var main = this.$slots.default;
            var footer = this.$slots.footer;
            console.log(footer);
            console.log(footer[0].elm.innerHTML);
        }
    });

    var app = new Vue({
        el: '#app'
    })
</script>
```

通过$slots 可以访问某个具名 slot，this.$slots.default 包括了所有没有被包含在具名 slot 中的节点。尝试编写代码，查看两个 console 打印的内容。

$slots 在业务中几乎用不到，在用 render 函数（进阶篇中将介绍）创建组件时会比较有用，但主要还是用于独立组件开发中。

7.5 组件高级用法

本节会介绍组件的一些高级用法，这些用法在实际业务中不是很常用，但在独立组件开发时可能会用到。如果你感觉以上内容已经足够完成你的业务开发了，可以跳过本节；如果你想继续探索 Vue 组件的奥秘，读完本节会对你有很大的启发。

7.5.1 递归组件

组件在它的模板内可以递归地调用自己，只要给组件设置 name 的选项就可以了。示例代码如下：

```
<div id="app">
    <child-component :count="1"></child-component>
</div>
<script>
    Vue.component('child-component', {
        name: 'child-component',
        props: {
            count: {
                type: Number,
                default: 1
            }
        },
        template: '\
        <div class="child">\
            <child-component\
                :count="count + 1"\
                v-if="count < 3"></child-component>\
        </div>',
    });

    var app = new Vue({
        el: '#app'
    })
</script>
```

设置 name 后，在组件模板内就可以递归使用了，不过需要注意的是，必须给一个条件来限制递归数量，否则会抛出错误：max stack size exceeded。

组件递归使用可以用来开发一些具有未知层级关系的独立组件，比如级联选择器和树形控件等，如图 7-4 和图 7-5 所示。

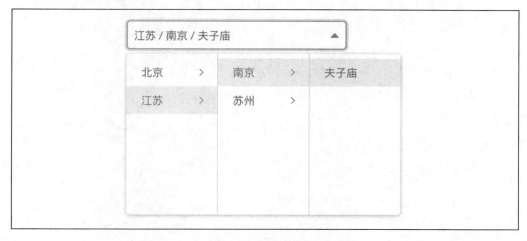

图 7-4 级联选择器

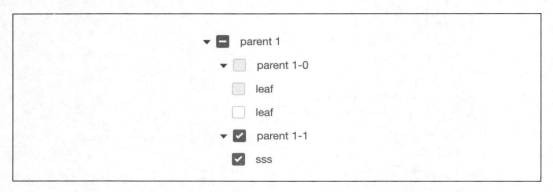

图 7-5　树形控件

在实战篇里，我们会详细介绍级联选择器的实现。

7.5.2　内联模板

组件的模板一般都是在 template 选项内定义的，Vue 提供了一个内联模板的功能，在使用组件时，给组件标签使用 inline-template 特性，组件就会把它的内容当作模板，而不是把它当内容分发，这让模板更灵活。示例代码如下：

```
<div id="app">
    <child-component inline-template>
        <div>
            <h2>在父组件中定义子组件的模板</h2>
            <p>{{ message }}</p>
            <p>{{ msg }}</p>
        </div>
    </child-component>
</div>
<script>
    Vue.component('child-component', {
        data: function () {
            return {
                msg: '在子组件声明的数据'
            }
        }
    });

    var app = new Vue({
        el: '#app',
        data: {
            message: '在父组件声明的数据'
        }
    })
</script>
```

渲染后的结果为：

```
<div id="app">
    <div>
        <h2>在父组件中定义子组件的模板</h2>
        <p>在父组件声明的数据</p>
        <p>在子组件声明的数据</p>
    </div>
</div>
```

在父组件中声明的数据 message 和子组件中声明的数据 msg，两个都可以渲染（如果同名，优先使用子组件的数据）。这反而是内联模板的缺点，就是作用域比较难理解，如果不是非常特殊的场景，建议不要轻易使用内联模板。

7.5.3 动态组件

Vue.js 提供了一个特殊的元素<component> 用来动态地挂载不同的组件，使用 is 特性来选择要挂载的组件。示例代码如下：

```
<div id="app">
    <component :is="currentView"></component>
    <button @click="handleChangeView('A')">切换到 A</button>
    <button @click="handleChangeView('B')">切换到 B</button>
    <button @click="handleChangeView('C')">切换到 C</button>
</div>
<script>
    var app = new Vue({
        el: '#app',
        components: {
            comA: {
                template: '<div>组件 A</div>'
            },
            comB: {
                template: '<div>组件 B</div>'
            },
            comC: {
                template: '<div>组件 C</div>'
            }
        },
        data: {
            currentView: 'comA'
        },
        methods: {
            handleChangeView: function (component) {
```

```
            this.currentView = 'com' + component;
        }
    }
})
</script>
```

动态地改变 currentView 的值就可以动态挂载组件了。也可以直接绑定在组件对象上：

```
<div id="app">
    <component :is="currentView"></component>
</div>
<script>
    var Home = {
        template: '<p>Welcome home!</p>'
    };
    var app = new Vue({
        el: '#app',
        data: {
            currentView: Home
        }
    })
</script>
```

7.5.4 异步组件

当你的工程足够大，使用的组件足够多时，是时候考虑下性能问题了，因为一开始把所有的组件都加载是没必要的一笔开销。好在 Vue.js 允许将组件定义为一个工厂函数，动态地解析组件。Vue.js 只在组件需要渲染时触发工厂函数，并且把结果缓存起来，用于后面的再次渲染。例如下面的示例：

```
<div id="app">
    <child-component></child-component>
</div>
<script>
    Vue.component('child-component', function (resolve, reject) {
        window.setTimeout(function () {
            resolve({
                template: '<div>我是异步渲染的</div>'
            });
        }, 2000);
    });
    var app = new Vue({
        el: '#app'
    })
</script>
```

工厂函数接收一个 resolve 回调，在收到从服务器下载的组件定义时调用。也可以调用 reject(reason)指示加载失败。这里 setTimeout 只是为了演示异步，具体的下载逻辑可以自己决定，比如把组件配置写成一个对象配置，通过 Ajax 来请求，然后调用 resolve 传入配置选项。

在进阶篇里，我们还会介绍主流的打包编译工具 webpack 和.vue 单文件的用法，更优雅地实现异步组件（路由）。

7.6 其 他

7.6.1 $nextTick

我们先来看这样一个场景：有一个 div，默认用 v-if 将它隐藏，点击一个按钮后，改变 v-if 的值，让它显示出来，同时拿到这个 div 的文本内容。如果 v-if 的值是 false，直接去获取 div 的内容是获取不到的，因为此时 div 还没有被创建出来，那么应该在点击按钮后，改变 v-if 的值为 true，div 才会被创建，此时再去获取，示例代码如下：

```
<div id="app">
    <div id="div" v-if="showDiv">这是一段文本</div>
    <button @click="getText">获取 div 内容</button>
</div>
<script>
    var app = new Vue({
        el: '#app',
        data: {
            showDiv: false
        },
        methods: {
            getText: function () {
                this.showDiv = true;
                var text = document.getElementById('div').innerHTML;
                console.log(text);
            }
        }
    })
</script>
```

这段代码并不难理解，但是运行后在控制台会抛出一个错误：Cannot read property 'innerHTML' of null，意思就是获取不到 div 元素。这里就涉及 Vue 一个重要的概念：异步更新队列。

Vue 在观察到数据变化时并不是直接更新 DOM，而是开启一个队列，并缓冲在同一事件循环中发生的所有数据改变。在缓冲时会去除重复数据，从而避免不必要的计算和 DOM 操作。然后，在下一个事件循环 tick 中，Vue 刷新队列并执行实际（已去重的）工作。所以如果你用一个 for 循环来动态改变数据 100 次，其实它只会应用最后一次改变，如果没有这种机制，DOM 就要重绘 100 次，这固然是一个很大的开销。

Vue 会根据当前浏览器环境优先使用原生的 Promise.then 和 MutationObserver，如果都不支持，就会采用 setTimeout 代替。

知道了 Vue 异步更新 DOM 的原理，上面示例的报错也就不难理解了。事实上，在执行 this.showDiv = true;时，div 仍然还是没有被创建出来，直到下一个 Vue 事件循环时，才开始创建。$nextTick 就是用来知道什么时候 DOM 更新完成的，所以上面的示例代码需要修改为：

```html
<div id="app">
    <div id="div" v-if="showDiv">这是一段文本</div>
    <button @click="getText">获取 div 内容</button>
</div>
<script>
    var app = new Vue({
        el: '#app',
        data: {
            showDiv: false
        },
        methods: {
            getText: function () {
                this.showDiv = true;
                this.$nextTick(function () {
                    var text = document.getElementById('div').innerHTML;
                    console.log(text);
                });
            }
        }
    })
</script>
```

这时再点击按钮，控制台就打印出 div 的内容"这是一段文本"了。

理论上，我们应该不用去主动操作 DOM，因为 Vue 的核心思想就是数据驱动 DOM，但在很多业务里，我们避免不了会使用一些第三方库，比如 popper.js（https://popper.js.org/）、swiper（http://idangero.us/swiper/）等，这些基于原生 JavaScript 的库都有创建和更新及销毁的完整生命周期，与 Vue 配合使用时，就要利用好$nextTick 。

7.6.2 X-Templates

如果你没有使用 webpack、gulp 等工具，试想一下你的组件 template 的内容很冗长、复杂，如果都在 JavaScript 里拼接字符串，效率是很低的，因为不能像写 HTML 那样舒服。Vue 提供了另一种定义模板的方式，在<script> 标签里使用 text/x-template 类型，并且指定一个 id，将这个 id 赋给 template。示例代码如下：

```html
<div id="app">
    <my-component></my-component>
    <script type="text/x-template" id="my-component">
```

```
        <div>这是组件的内容</div>
    </script>
</div>
<script>
    Vue.component('my-component', {
        template: '#my-component'
    });

    var app = new Vue({
        el: '#app'
    })
</script>
```

在<script> 标签里，你可以愉快地写 HTML 代码，不用考虑换行等问题。

很多刚接触 Vue 开发的新手会非常喜欢这个功能，因为用它，再加上组件知识，就可以很轻松地完成交互相对复杂的页面和应用了。如果再配合一些构建工具（gulp）组织好代码结构，开发一些中小型产品是没有问题的。不过，Vue 的初衷并不是滥用它，因为它将模板和组件的其他定义隔离了。在进阶篇里，我们会介绍如何使用 webpack 来编译 .vue 的单文件，从而优雅地解决 HTML 书写的问题。

7.6.3 手动挂载实例

我们现在所创建的实例都是通过 new Vue()的形式创建出来的。在一些非常特殊的情况下，我们需要动态地去创建 Vue 实例，Vue 提供了 Vue.extend 和$mount 两个方法来手动挂载一个实例。

Vue.extend 是基础 Vue 构造器，创建一个"子类"，参数是一个包含组件选项的对象。

如果 Vue 实例在实例化时没有收到 el 选项，它就处于"未挂载"状态，没有关联的 DOM 元素。可以使用$mount()手动地挂载一个未挂载的实例。这个方法返回实例自身，因而可以链式调用其他实例方法。示例代码如下：

```
<div id="mount-div">

</div>
<script>
    var MyComponent = Vue.extend({
        template: '<div>Hello: {{ name }}</div>',
        data: function () {
            return {
                name: 'Aresn'
            }
        }
    });

    new MyComponent().$mount('#mount-div');
</script>
```

运行后，id 为 mount-div 的 div 元素会被替换为组件 MyComponent 的 template 的内容：

```
<div>Hello: Aresn</div>
```

除了这种写法外，以下两种写法也是可以的：

```
new MyComponent().$mount('#mount-div');
// 同上
new MyComponent({
    el: '#mount-div'
});
//或者，在文档之外渲染并且随后挂载
var component = new MyComponent().$mount();
document.getElementById('mount-div').appendChild(component.$el);
```

手动挂载实例（组件）是一种比较极端的高级用法，在业务中几乎用不到，只在开发一些复杂的独立组件时可能会使用，所以只做了解就好。

7.7 实战：两个常用组件的开发

本节以组件知识为基础，整合指令、事件等前面章节的内容，开发两个业务中常用的组件，即数字输入框和标签页。

7.7.1 开发一个数字输入框组件

数字输入框是对普通输入框的扩展，用来快捷输入一个标准的数字，如图 7-6 所示。

图 7-6 数字输入框

数字输入框只能输入数字，而且有两个快捷按钮，可以直接减 1 或加 1。除此之外，还可以设置初始值、最大值、最小值，在数值改变时，触发一个自定义事件来通知父组件。

了解了基本需求后，我们先定义目录文件：

- index.html 入口页
- input-number.js 数字输入框组件
- index.js 根实例

因为该示例是以交互功能为主，所以就不写 CSS 美化样式了。

首先写入基本的结构代码，初始化项目。

index.html：

```html
<!DOCTYPE html>
<html>
<head>
    <meta charset="utf-8">
    <title>数字输入框组件</title>
</head>
<body>
    <div id="app">

    </div>
    <script src="https://unpkg.com/vue/dist/vue.min.js"></script>
    <script src="input-number.js"></script>
    <script src="index.js"></script>
</body>
</html>
```

index.js：

```js
var app = new Vue({
    el: '#app'
});
```

input-number.js：

```js
Vue.component('input-number', {
    template: '\
        <div class="input-number"> \
            \
        </div>',
    props: {
        max: {
            type: Number,
            default: Infinity
        },
        min: {
            type: Number,
            default: -Infinity
        },
        value: {
            type: Number,
            default: 0
        }
    }
});
```

该示例的主角是 input-number.js，所有的组件配置都在这里面定义。先在 template 里定义了组件的根节点，因为是独立组件，所以应该对每个 prop 进行校验。这里根据需求有最大值、最小值、默认值（也就是绑定值）3 个 prop，max 和 min 都是数字类型，默认值是正无限大和负无限大；value 也是数字类型，默认值是 0。

接下来，我们先在父组件引入 input-number 组件，并给它一个默认值 5，最大值 10，最小值 0。

index.js：

```
var app = new Vue({
    el: '#app',
    data: {
        value: 5
    }
});
```

index.html：

```
<div id="app">
    <input-number v-model="value" :max="10" :min="0"></input-number>
</div>
```

value 是一个关键的绑定值，所以用了 v-model，这样既优雅地实现了双向绑定，也让 API 看起来很合理。大多数的表单类组件都应该有一个 v-model，比如输入框、单选框、多选框、下拉选择器等。

剩余的代码量就都聚焦到了 input-number.js 上。

我们之前介绍过，Vue 组件是单向数据流，所以无法从组件内部直接修改 prop value 的值。解决办法也介绍过，就是给组件声明一个 data，默认引用 value 的值，然后在组件内部维护这个 data：

```
Vue.component('input-number', {
    // ...
    data: function () {
        return {
            currentValue: this.value
        }
    }
})
```

这样只解决了初始化时引用父组件 value 的问题，但是如果从父组件修改了 value，input-number 组件的 currentValue 也要一起更新。为了实现这个功能，我们需要用到一个新的概念：监听（watch）。watch 选项用来监听某个 prop 或 data 的改变，当它们发生变化时，就会触发 watch 配置的函数，从而完成我们的业务逻辑。在本例中，我们要监听两个量：value 和 currentValue。监听 value 是要知晓从父组件修改了 value，监听 currentValue 是为了当 currentValue 改变时，更新 value。相关代码如下：

```
Vue.component('input-number', {
    // ...
    data: function () {
        return {
            currentValue: this.value
        }
    },
    watch: {
        currentValue: function (val) {
            this.$emit('input', val);
            this.$emit('on-change', val);
        },
        value: function (val) {
            this.updateValue(val);
        }
    },
    methods: {
        updateValue: function (val) {
            if (val > this.max) val = this.max;
            if (val < this.min) val = this.min;
            this.currentValue = val;
        }
    },
    mounted: function () {
        this.updateValue(this.value);
    }
})
```

从父组件传递过来的 value 有可能是不符合当前条件的（大于 max，或小于 min），所以在选项 methods 里写了一个方法 updateValue，用来过滤出一个正确的 currentValue。

watch 监听的数据的回调函数有 2 个参数可用，第一个是新的值，第二个是旧的值，这里没有太复杂的逻辑，就只用了第一个参数。在回调函数里，this 是指向当前组件实例的，所以可以直接调用 this.updateValue()，因为 Vue 代理了 props、data、computed 及 methods。

监听 currentValue 的回调里，this.$emit('input', val)是在使用 v-model 时改变 value 的；this.$emit('on-change', val)是触发自定义事件 on-change，用于告知父组件数字输入框的值有所改变（示例中没有使用该事件）。

在生命周期 mounted 钩子里也调用了 updateValue()方法，是因为第一次初始化时，也对 value 进行了过滤。这里也有另一种写法，在 data 选项返回对象前进行过滤：

```
Vue.component('input-number', {
    // ...
    data: function () {
        var val = this.value;
```

```
        if (val > this.max) val = this.max;
        if (val < this.min) val = this.min;

        return {
            currentValue: val
        }
    }
});
```

实现的效果是一样的。

最后剩余的就是补全模板 template，内容是一个输入框和两个按钮，相关代码如下：

```
function isValueNumber (value) {
    return (/(^-?[0-9]+\.{1}\d+$)|(^-?[1-9][0-9]*$)|(^-?0{1}$)/).test(value + '');
}

Vue.component('input-number', {
    // ...
    template: '\
        <div class="input-number"> \
            <input \
                type="text" \
                :value="currentValue" \
                @change="handleChange"> \
            <button \
                @click="handleDown" \
                :disabled="currentValue <= min">-</button> \
            <button \
                @click="handleUp" \
                :disabled="currentValue >= max">+</button> \
        </div>',
    methods: {
        handleDown: function () {
            if (this.currentValue <= this.min) return;
            this.currentValue -= 1;
        },
        handleUp: function () {
            if (this.currentValue >= this.max) return;
            this.currentValue += 1;
        },
        handleChange: function (event) {
            var val = event.target.value.trim();
```

```
                var max = this.max;
                var min = this.min;

                if (isValueNumber(val)) {
                    val = Number(val);
                    this.currentValue = val;

                    if (val > max) {
                        this.currentValue = max;
                    } else if (val < min) {
                        this.currentValue = min;
                    }
                } else {
                    event.target.value = this.currentValue;
                }
            }
        }
    });
```

input 绑定了数据 currentValue 和原生的 change 事件，在句柄 handleChange 函数中，判断了当前输入的是否是数字。注意，这里绑定的 currentValue 也是单向数据流，并没有用 v-model，所以在输入时，currentValue 的值并没有实时改变。如果输入的不是数字（比如英文和汉字等），就将输入的内容重置为之前的 currentValue。如果输入的是符合要求的数字，就把输入的值赋给 currentValue。

数字输入框组件的核心逻辑就是这些。回顾一下我们设计一个通用组件的思路，首先，在写代码前一定要明确需求，然后规划好 API。一个 Vue 组件的 API 只来自 props、events 和 slots，确定好这 3 部分的命名、规则，剩下的逻辑即使第一版没有做好，后续也可以迭代完善。但是 API 如果没有设计好，后续再改对使用者成本就很大了。

完整的示例代码如下：

index.html

```html
<!DOCTYPE html>
<html>
<head>
    <meta charset="utf-8">
    <title>数字输入框组件</title>
</head>
<body>
    <div id="app">
        <input-number v-model="value" :max="10" :min="0"></input-number>
    </div>
    <script src="https://unpkg.com/vue/dist/vue.min.js"></script>
```

```html
        <script src="input-number.js"></script>
        <script src="index.js"></script>
    </body>
</html>
```

index.js

```js
var app = new Vue({
    el: '#app',
    data: {
        value: 5
    }
});
```

input-number.js

```js
function isValueNumber (value) {
    return (/(^-?[0-9]+\.{1}\d+$)|(^-?[1-9][0-9]*$)|(^-?0{1}$)/).test(value + '');
}

Vue.component('input-number', {
    template: '\
        <div class="input-number"> \
            <input \
                type="text" \
                :value="currentValue" \
                @change="handleChange"> \
            <button \
                @click="handleDown" \
                :disabled="currentValue <= min">-</button> \
            <button \
                @click="handleUp" \
                :disabled="currentValue >= max">+</button> \
        </div>',
    props: {
        max: {
            type: Number,
            default: Infinity
        },
        min: {
            type: Number,
            default: -Infinity
        },
        value: {
```

```js
            type: Number,
            default: 0
        }
    },
    data: function () {
        return {
            currentValue: this.value
        }
    },
    watch: {
        currentValue: function (val) {
            this.$emit('input', val);
            this.$emit('on-change', val);
        },
        value: function (val) {
            this.updateValue(val);
        }
    },
    methods: {
        handleDown: function () {
            if (this.currentValue <= this.min) return;
            this.currentValue -= 1;
        },
        handleUp: function () {
            if (this.currentValue >= this.max) return;
            this.currentValue += 1;
        },
        updateValue: function (val) {
            if (val > this.max) val = this.max;
            if (val < this.min) val = this.min;
            this.currentValue = val;
        },
        handleChange: function (event) {
            var val = event.target.value.trim();

            var max = this.max;
            var min = this.min;

            if (isValueNumber(val)) {
                val = Number(val);
                this.currentValue = val;

                if (val > max) {
```

```
                this.currentValue = max;
            } else if (val < min) {
                this.currentValue = min;
            }
        } else {
            event.target.value = this.currentValue;
        }
    }
},
mounted: function () {
    this.updateValue(this.value);
}
});
```

练习 1：在输入框聚焦时，增加对键盘上下按键的支持，相当于加 1 和减 1。

练习 2：增加一个控制步伐的 prop —— step，比如设置为 10，点击加号按钮，一次增加 10。

7.7.2 开发一个标签页组件

本小节将开发一个比较有挑战的组件：标签页组件。标签页（即选项卡切换组件）是网页布局中经常用到的元素，常用于平级区域大块内容的收纳和展现，如图 7-7 所示。

图 7-7 标签页

根据上个示例的经验，我们先分析业务需求，制定出 API，这样不至于一上来就无从下手。

每个标签页的主体内容肯定是由使用组件的父级控制的，所以这部分是一个 slot，而且 slot 的数量决定了标签切换按钮的数量。假设我们有 3 个标签页，点击每个标签按钮时，另外的两个标签对应的 slot 应该被隐藏掉。一般这个时候，比较容易想到的解决办法是，在 slot 里写 3 个 div，在接收到切换通知时，显示和隐藏相关的 div。这样设计没有问题，只不过体现不出组件的价值来，因为我们还是写了一些与业务无关的交互逻辑，而这部分逻辑最好组件本身帮忙处理了，我们只用聚焦在 slot 内容本身，这才是与我们业务最相关的。这种情况下，我们再定义一个子组件 pane，嵌套在标签页组件 tabs 里，我们的业务代码都放在 pane 的 slot 内，而 3 个 pane 组件作为整体成为 tabs 的 slot。

由于 tabs 和 pane 两个组件是分离的，但是 tabs 组件上的标题应该由 pane 组件来定义，因为 slot 是写在 pane 里，因此在组件初始化（及标签标题动态改变）时，tabs 要从 pane 里获取标题，并保存起来，自己使用。

确定好了结构，我们先创建所需的文件：

- index.html 入口页
- style.css 样式表
- tabs.js 标签页外层的组件 tabs
- pane.js 标签页嵌套的组件 pane

先初始化各个文件。

index.html：

```
<!DOCTYPE html>
<html>
<head>
    <meta charset="utf-8">
    <title>标签页组件</title>
    <link rel="stylesheet" type="text/css" href="style.css">
</head>
<body>
    <div id="app" v-cloak>

    </div>
    <script src="https://unpkg.com/vue/dist/vue.min.js"></script>
    <script src="pane.js"></script>
    <script src="tabs.js"></script>
    <script type="text/javascript">
        var app = new Vue({
            el: '#app'
        })
    </script>
</body>
</html>
```

tabs.js：

```
Vue.component('tabs', {
    template: '\
        <div class="tabs"> \
            <div class="tabs-bar"> \
                <!-- 标签页标题，这里要用 v-for --> \
            </div> \
            <div class="tabs-content"> \
                <!-- 这里的 slot 就是嵌套的 pane --> \
                <slot></slot> \
            </div> \
```

```
        </div>'
})
```

pane.js：

```
Vue.component('pane', {
    name: 'pane',
    template: '\
        <div class="pane"> \
            <slot></slot> \
        </div>'
})
```

pane 需要控制标签页内容的显示与隐藏，设置一个 data: show，并且用 v-show 指令来控制元素：

```
template: '\
    <div class="pane" v-show="show"> \
        <slot></slot> \
    </div>',
data: function () {
    return {
        show: true
    }
}
```

当点击到这个 pane 对应的标签页标题按钮时，此 pane 的 show 值设置为 true，否则应该是 false，这步操作是在 tabs 组件完成的，我们稍后再介绍。

既然要点击对应的标签页标题按钮，那应该有一个唯一的值来标识这个 pane，我们可以设置一个 prop：name 让用户来设置，但它不是必需的，如果使用者不设置，可以默认从 0 开始自动设置，这步操作仍然是 tabs 执行的，因为 pane 本身并不知道自己是第几个。除了 name，还需要标签页标题的 prop：label，tabs 组件需要将它显示在标签页标题里。这部分代码如下：

```
props: {
    name: {
        type: String
    },
    label: {
        type: String,
        default: ''
    }
}
```

上面的 prop：label 用户是可以动态调整的，所以在 pane 初始化及 label 更新时，都要通知父组件也更新，因为是独立组件，所以不能依赖像 bus.js 或 vuex 这样的状态管理办法，我们可以直接通过 this.$parent 访问 tabs 组件的实例来调用它的方法更新标题，该方法名暂定为 updateNav。注

意,在业务中尽可能不要使用$parent 来操作父链,这种方法适合于标签页这样的独立组件。这部分代码如下:

```
methods: {
    updateNav () {
        this.$parent.updateNav();
    }
},
watch: {
    label () {
        this.updateNav();
    }
},
mounted () {
    this.updateNav();
}
```

在生命周期 mounted,也就是 pane 初始化时,调用一遍 tabs 的 updateNav 方法,同时监听了 prop:label,在 label 更新时,同样调用。

剩余任务就是完成 tabs.js 组件。

首先需要把 pane 组件设置的标题动态渲染出来,也就是当 pane 触发 tabs 的 updateNav 方法时,更新标题内容。我们先看一下这部分的代码:

```
Vue.component('tabs', {
    // ...
    data: function () {
        return {
            // 用于渲染 tabs 的标题
            navList: []
        }
    },
    methods: {
        getTabs () {
            // 通过遍历子组件,得到所有的 pane 组件
            return this.$children.filter(function (item) {
                return item.$options.name === 'pane';
            });
        },
        updateNav () {
            this.navList = [];
            // 设置对 this 的引用,在 function 回调里,this 指向的并不是 Vue 实例
            var _this = this;

            this.getTabs().forEach(function (pane, index) {
```

```javascript
            _this.navList.push({
                label: pane.label,
                name: pane.name || index
            });
            // 如果没有给 pane 设置 name，默认设置它的索引
            if (!pane.name) pane.name = index;
            // 设置当前选中的 tab 的索引，在后面介绍
            if (index === 0) {
                if (!_this.currentValue) {
                    _this.currentValue = pane.name || index;
                }
            }
        });

        this.updateStatus();
    },
    updateStatus () {
        var tabs = this.getTabs();
        var _this = this;
        // 显示当前选中的 tab 对应的 pane 组件，隐藏没有选中的
        tabs.forEach(function (tab) {
            return tab.show = tab.name === _this.currentValue;
        })
    }
}
})
```

getTabs 是一个公用的方法，使用 this.$children 来拿到所有的 pane 组件实例。

需要注意的是，在 methods 里使用了有 function 回调的方法时（例如遍历数组的方法 forEach），在回调内的 this 不再执行当前的 Vue 实例，也就是 tabs 组件本身，所以要在外层设置一个 _this = this 的局部变量来间接使用 this。如果你熟悉 ES2015，也可以直接使用箭头函数=>，我们会在实战篇里介绍相关的用法。

遍历了每一个 pane 组件后，把它的 label 和 name 提取出来，构成一个 Object 并添加到数据 navList 数组里，后面我们会在 template 里用到它。

设置完 navList 数组后，我们调用了 updateStatus 方法，又将 pane 组件遍历了一遍，不过这时是为了将当前选中的 tab 对应的 pane 组件内容显示出来，把没有选中的隐藏掉。因为在上一步操作里，我们有可能需要设置 currentValue 来标识当前选中项的 name（在用户没有设置 value 时，才会自动设置），所以必须要遍历 2 次才可以。

拿到 navList 后，就需要对它用 v-for 指令把 tab 的标题渲染出来，并且判断每个 tab 当前的状态。这部分代码如下：

```javascript
Vue.component('tabs', {
    template: '\
```

```
            <div class="tabs"> \
                <div class="tabs-bar"> \
                    <div \
                        :class="tabCls(item)" \
                        v-for="(item, index) in navList" \
                        @click="handleChange(index)"> \
                        {{ item.label }} \
                    </div> \
                </div> \
                <div class="tabs-content"> \
                    <slot></slot> \
                </div> \
            </div>',
    props: {
        // 这里的 value 是为了可以使用 v-model
        value: {
            type: [String, Number]
        }
    },
    data: function () {
        return {
            // 因为不能修改 value,所以复制一份自己维护
            currentValue: this.value,
            navList: []
        }
    },
    methods: {
        tabCls: function (item) {
            return [
                'tabs-tab',
                {
                    //给当前选中的 tab 加一个 class
                    'tabs-tab-active': item.name === this.currentValue
                }
            ]
        },
        // 点击 tab 标题时触发
        handleChange: function (index) {
            var nav = this.navList[index];
            var name = nav.name;
            // 改变当前选中的 tab ,并触发下面的 watch
            this.currentValue = name;
            // 更新 value
```

```
            this.$emit('input', name);
            // 触发一个自定义事件，供父级使用
            this.$emit('on-click', name);
        }
    },
    watch: {
        value: function (val) {
            this.currentValue = val;
        },
        currentValue: function () {
            //在当前选中的 tab 发生变化时，更新 pane 的显示状态
            this.updateStatus();
        }
    }
})
```

在使用 v-for 指令循环显示 tab 标题时，使用 v-bind:class 指向了一个名为 tabCls 的 methods 来动态设置 class 名称。因为计算属性不能接收参数，无法知道当前 tab 是否是选中的，所以这里我们才用到 methods，不过要知道，methods 是不缓存的，可以回顾关于计算属性的章节。

点击每个 tab 标题时，会触发 handleChange 方法来改变当前选中 tab 的索引，也就是 pane 组件的 name。在 watch 选项里，我们监听了 currentValue，当其发生变化时，触发 updateStatus 方法来更新 pane 组件的显示状态。

以上就是标签页组件的核心代码分解。总结一下该示例的技术难点：使用了组件嵌套的方式，将一系列 pane 组件作为 tabs 组件的 slot；tabs 组件和 pane 组件通信上，使用了 $parent 和 $children 的方法访问父链和子链；定义了 prop: value 和 data: currentValue，使用 $emit('input') 来实现 v-model 的用法。

以下是标签页组件的完整代码。

index.html：

```html
<!DOCTYPE html>
<html>
<head>
    <meta charset="utf-8">
    <title>标签页组件</title>
    <link rel="stylesheet" type="text/css" href="style.css">
</head>
<body>
    <div id="app" v-cloak>
        <tabs v-model="activeKey">
            <pane label="标签一" name="1">
                标签一的内容
            </pane>
```

```html
            <pane label="标签二" name="2">
                标签二的内容
            </pane>
            <pane label="标签三" name="3">
                标签三的内容
            </pane>
        </tabs>
    </div>
    <script src="https://unpkg.com/vue/dist/vue.min.js"></script>
    <script src="pane.js"></script>
    <script src="tabs.js"></script>
    <script type="text/javascript">
        var app = new Vue({
            el: '#app',
            data: {
                activeKey: '1'
            }
        })
    </script>
</body>
</html>
```

pane.js：

```javascript
Vue.component('pane', {
    name: 'pane',
    template: '\
        <div class="pane" v-show="show"> \
            <slot></slot> \
        </div>',
    props: {
        name: {
            type: String
        },
        label: {
            type: String,
            default: ''
        }
    },
    data: function () {
        return {
            show: true
        }
    },
```

```
    methods: {
        updateNav () {
            this.$parent.updateNav();
        }
    },
    watch: {
        label () {
            this.updateNav();
        }
    },
    mounted () {
        this.updateNav();
    }
})
```

tabs.js：

```
Vue.component('tabs', {
    template: '\
        <div class="tabs"> \
            <div class="tabs-bar"> \
                <div \
                    :class="tabCls(item)" \
                    v-for="(item, index) in navList" \
                    @click="handleChange(index)"> \
                    {{ item.label }} \
                </div> \
            </div> \
            <div class="tabs-content"> \
                <slot></slot> \
            </div> \
        </div>',
    props: {
        value: {
            type: [String, Number]
        }
    },
    data: function () {
        return {
            currentValue: this.value,
            navList: []
        }
    },
    methods: {
```

```js
        tabCls: function (item) {
            return [
                'tabs-tab',
                {
                    'tabs-tab-active': item.name === this.currentValue
                }
            ]
        },
        getTabs () {
            return this.$children.filter(function (item) {
                return item.$options.name === 'pane';
            });
        },
        updateNav () {
            this.navList = [];
            var _this = this;

            this.getTabs().forEach(function (pane, index) {
                _this.navList.push({
                    label: pane.label,
                    name: pane.name || index
                });
                if (!pane.name) pane.name = index;
                if (index === 0) {
                    if (!_this.currentValue) {
                        _this.currentValue = pane.name || index;
                    }
                }
            });

            this.updateStatus();
        },
        updateStatus () {
            var tabs = this.getTabs();
            var _this = this;

            tabs.forEach(function (tab) {
                return tab.show = tab.name === _this.currentValue;
            })
        },
        handleChange: function (index) {
            var nav = this.navList[index];
            var name = nav.name;
```

```js
                this.currentValue = name;
                this.$emit('input', name);
                this.$emit('on-click', name);
            }
        },
        watch: {
            value: function (val) {
                this.currentValue = val;
            },
            currentValue: function () {
                this.updateStatus();
            }
        }
    })
```

style.css：

```css
[v-cloak] {
    display: none;
}
.tabs{
    font-size: 14px;
    color: #657180;
}
.tabs-bar:after{
    content: '';
    display: block;
    width: 100%;
    height: 1px;
    background: #d7dde4;
    margin-top: -1px;
}
.tabs-tab{
    display: inline-block;
    padding: 4px 16px;
    margin-right: 6px;
    background: #fff;
    border: 1px solid #d7dde4;
    cursor: pointer;
    position: relative;
}
.tabs-tab-active{
    color: #3399ff;
```

```
        border-top: 1px solid #3399ff;
        border-bottom: 1px solid #fff;
    }
    .tabs-tab-active:before{
        content: '';
        display: block;
        height: 1px;
        background: #3399ff;
        position: absolute;
        top: 0;
        left: 0;
        right: 0;
    }
    .tabs-content{
        padding: 8px 0;
    }
```

练习 1：给 pane 组件新增一个 prop：closable 的布尔值，来支持是否可以关闭这个 pane，如果开启，在 tabs 的标签标题上会有一个关闭的按钮。

在初始化 pane 时，我们是在 mounted 里通知的，关闭时，你会用到 beforeDestroy。

练习 2：尝试在切换 pane 的显示与隐藏时，使用滑动的动画。提示：可以使用 CSS 3 的 transform: translateX。

第 8 章

自定义指令

在第 5 章里我们已经介绍过了许多 Vue 内置的指令，比如 v-if、v-show 等，这些丰富的内置指令能满足我们的绝大部分业务需求，不过在需要一些特殊功能时，我们仍然希望对 DOM 进行底层的操作，这时就要用到自定义指令。

8.1 基本用法

自定义指令的注册方法和组件很像，也分全局注册和局部注册，比如注册一个 v-focus 的指令，用于在 <input>、<textarea> 元素初始化时自动获得焦点，两种写法分别是：

```
// 全局注册
Vue.directive('focus', {
    //指令选项
});

// 局部注册
var app = new Vue({
    el: '#app',
    directives: {
        focus: {
            //指令选项
        }
    }
})
```

写法与组件基本类似，只是方法名由 component 改为了 directive。上例只是注册了自定义指令 v-focus，还没有实现具体功能，下面具体介绍自定义指令的各个选项。

自定义指令的选项是由几个钩子函数组成的，每个都是可选的。

- bind：只调用一次，指令第一次绑定到元素时调用，用这个钩子函数可以定义一个在绑定时执行一次的初始化动作。
- inserted：被绑定元素插入父节点时调用（父节点存在即可调用，不必存在于 document 中）。
- update：被绑定元素所在的模板更新时调用，而不论绑定值是否变化。通过比较更新前后的绑定值，可以忽略不必要的模板更新。
- componentUpdated：被绑定元素所在模板完成一次更新周期时调用。
- unbind：只调用一次，指令与元素解绑时调用。

可以根据需求在不同的钩子函数内完成逻辑代码，例如上面的 v-focus，我们希望在元素插入父节点时就调用，那用到的最好是 inserted。示例代码如下：

```html
<div id="app">
    <input type="text" v-focus>
</div>
<script>
    Vue.directive('focus', {
        inserted: function (el) {
            // 聚焦元素
            el.focus();
        }
    });

    var app = new Vue({
        el: '#app'
    })
</script>
```

在浏览器中的效果如图 8-1 所示。

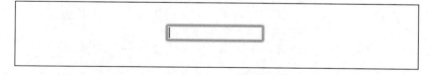

图 8-1　v-focus 渲染后的效果

可以看到，打开这个页面，input 输入框就自动获得了焦点，成为可输入状态。
每个钩子函数都有几个参数可用，比如上面我们用到了 el。它们的含义如下：

- el　指令所绑定的元素，可以用来直接操作 DOM。
- binding　一个对象，包含以下属性：

- ➢ name　指令名，不包括 v- 前缀。
- ➢ value　指令的绑定值，例如 v-my-directive="1 + 1"，value 的值是 2。
- ➢ oldValue　指令绑定的前一个值，仅在 update 和 componentUpdated 钩子中可用。无论值是否改变都可用。
- ➢ expression　绑定值的字符串形式。例如 v-my-directive="1 + 1", expression 的值是"1 + 1"。
- ➢ arg　传给指令的参数。例如 v-my-directive:foo，arg 的值是 foo。
- ➢ modifiers　一个包含修饰符的对象。例如 v-my-directive.foo.bar，修饰符对象 modifiers 的值是 { foo: true, bar: true }。

- vnode　Vue 编译生成的虚拟节点，在进阶篇中介绍。
- oldVnode　上一个虚拟节点仅在 update 和 componentUpdated 钩子中可用。

下面是结合了以上参数的一个具体示例，代码如下：

```
<div id="app">
    <div v-test:msg.a.b="message"></div>
</div>
<script>
    Vue.directive('test', {
        bind: function (el, binding, vnode) {
            var keys = [];
            for (var i in vnode) {
                keys.push(i);
            }
            el.innerHTML =
                'name: '       + binding.name + '<br>' +
                'value: '      + binding.value + '<br>' +
                'expression: ' + binding.expression + '<br>' +
                'argument: '   + binding.arg + '<br>' +
                'modifiers: '  + JSON.stringify(binding.modifiers) + '<br>' +
                'vnode keys: ' + keys.join(', ')
        }
    });

    var app = new Vue({
        el: '#app',
        data: {
            message: 'some text'
        }
    })
</script>
```

执行后，<div>的内容会使用 innerHTML 重置，结果为：

```
name: test
value: some text
expression: message
argument: msg
modifiers: {"a":true,"b":true}
vnode keys:
tag,data,children,text,elm,ns,context,functionalContext,key,componentOptions,componentInstance,parent,raw,isStatic,isRootInsert,isComment,isCloned,isOnce
```

在大多数使用场景，我们会在 bind 钩子里绑定一些事件，比如在 document 上用 addEventListener 绑定，在 unbind 里用 removeEventListener 解绑，比较典型的示例就是让这个元素随着鼠标拖拽。在后面的 8.2 节中，我们会详细介绍。

如果需要多个值，自定义指令也可以传入一个 JavaScript 对象字面量，只要是合法类型的 JavaScript 表达式都是可以的。示例代码如下：

```
<div id="app">
    <div v-test="{msg: 'hello', name: 'Aresn'}"></div>
</div>
<script>
    Vue.directive('test', {
        bind: function (el, binding, vnode) {
            console.log(binding.value.msg);
            console.log(binding.value.name);
        }
    });

    var app = new Vue({
        el: '#app'
    })
</script>
```

Vue 2.x 移除了大量 Vue 1.x 自定义指令的配置。在使用自定义指令时，应该充分理解业务需求，因为很多时候你需要的可能并不是自定义指令，而是组件。在下一节中，我们结合两个经典的示例来进一步了解自定义指令的使用场景和用法。

8.2 实　　战

8.2.1　开发一个可从外部关闭的下拉菜单

网页中有很多常见的下拉菜单，比如图 8-2 所示的用户信息菜单。

图 8-2 点击用户的下拉菜单

点击用户头像和名称,会弹出一个下拉菜单,然后点击页面中其他空白区域(除了菜单本身外),菜单就关闭了。本示例就用自定义指令来实现这样的需求。

先来分析一下如何实现。

该示例有两个特点,一是点击下拉菜单本身是不会关闭的,二是点击下拉菜单以外的所有区域都要关闭。点击所有区域可以在 document 上绑定 click 事件来实现,同时只要过滤出是否点击的是目标元素内部的元素即可。

示例最终呈现的效果如图 8-3 所示。

图 8-3 下拉菜单示例最终效果

首先初始化各个文件。

index.html :

```
<!DOCTYPE html>
<html>
<head>
    <meta charset="utf-8">
    <title>可从外部关闭的下拉菜单</title>
    <link rel="stylesheet" type="text/css" href="style.css">
```

```html
</head>
<body>
    <div id="app" v-cloak>

    </div>
    <script src="https://unpkg.com/vue/dist/vue.min.js"></script>
    <script src="clickoutside.js"></script>
    <script src="index.js"></script>
</body>
</html>
```

index.js：

```js
var app = new Vue({
    el: '#app'
});
```

clickoutside.js：

```js
Vue.directive('clickoutside', {

});
```

利用组件的基本知识很容易完成 index.html 和 index.js 的逻辑：

```html
<div id="app" v-cloak>
    <div class="main" v-clickoutside="handleClose">
        <button @click="show = !show">点击显示下拉菜单</button>
        <div class="dropdown" v-show="show">
            <p>下拉框的内容，点击外面区域可以关闭</p>
        </div>
    </div>
</div>
```

```js
var app = new Vue({
    el: '#app',
    data: {
        show: false
    },
    methods: {
        handleClose: function () {
            this.show = false;
        }
    }
});
```

逻辑很简单，点击按钮时显示 class 为 dropdown 的 div 元素。

自定义指令 v-clickoutside 绑定了一个函数 handleClose，用来关闭菜单。先来看一下 clickoutside.js 的内容：

```
Vue.directive('clickoutside', {
    bind: function (el, binding, vnode) {
        function documentHandler (e) {
            if (el.contains(e.target)) {
                return false;
            }
            if (binding.expression) {
                binding.value(e);
            }
        }
        el.__vueClickOutside__ = documentHandler;
        document.addEventListener('click', documentHandler);
    },
    unbind: function (el, binding) {
        document.removeEventListener('click', el.__vueClickOutside__);
        delete el.__vueClickOutside__;
    }
});
```

之前分析过，要在 document 上绑定 click 事件，所以在 bind 钩子内声明了一个函数 documentHandler，并将它作为句柄绑定在 document 的 click 事件上。documentHandler 函数做了两个判断，第一个是判断点击的区域是否是指令所在的元素内部，如果是，就跳出函数，不往下继续执行。

contains 方法是用来判断元素 A 是否包含了元素 B，包含返回 true，不包含返回 false。示例代码如下：

```
<!DOCTYPE html>
<html>
<head>
    <title>contains</title>
</head>
<body>
    <div id="parent">
        父元素
        <div id="children">子元素</div>
    </div>
    <script type="text/javascript">
        var A = document.getElementById('parent');
        var B = document.getElementById('children');
```

```
                console.log(A.contains(B));  // true
                console.log(B.contains(A));  // false
            </script>
        </body>
    </html>
```

第二个判断的是当前的指令 v-clickoutside 有没有写表达式,在该自定义指令中,表达式应该是一个函数,在过滤了内部元素后,点击外面任何区域应该执行用户表达式中的函数,所以 binding.value()就是用来执行当前上下文 methods 中指定的函数的。

与 Vue 1.x 不同的是,在自定义指令中,不能再用 this.xxx 的形式在上下文中声明一个变量,所以用了 el.__vueClickOutside__ 引用了 documentHandler,这样就可以在 unbind 钩子里移除对 document 的 click 事件监听。如果不移除,当组件或元素销毁时,它仍然存在于内存中。

以上代码分解与完整代码基本一致,不再重复提供。下面是 style.css 的代码:

```css
[v-cloak] {
    display: none;
}
.main{
    width: 125px;
}
button{
    display: block;
    width: 100%;
    color: #fff;
    background-color: #39f;
    border: 0;
    padding: 6px;
    text-align: center;
    font-size: 12px;
    border-radius: 4px;
    cursor: pointer;
    outline: none;
    position: relative;
}
button:active{
    top: 1px;
    left: 1px;
}
.dropdown{
    width: 100%;
    height: 150px;
    margin: 5px 0;
```

```
        font-size: 12px;
        background-color: #fff;
        border-radius: 4px;
        box-shadow: 0 1px 6px rgba(0,0,0,.2);
    }
    .dropdown p{
        display: inline-block;
        padding: 6px;
    }
```

练习 1：在 update 钩子中支持表达式的更新。

练习 2：扩展 clickoutside.js，实现在点击按钮显示下拉菜单后，通过按下键盘的 ESC 键也可以关闭下拉菜单。

练习 3：将练习 2 的 ESC 按键关闭功能作为可选项。提示，可以用修饰符，比如 v-clickoutside.esc。

8.2.2 开发一个实时时间转换指令 v-time

在一些社区，比如微博、朋友圈等，发布的动态会有一个相对本机时间转换后的相对时间，如图 8-4 中波浪线圈出的时间。

图 8-4　程序员社区 talkingcoder.com 最新文章列表

一般在服务端的存储时间格式是 Unix 时间戳，比如 2017-01-01 00:00:00 的时间戳是 1483200000。前端在拿到数据后，将它转换为可读的时间格式再显示出来。为了显示出实时性，在一些社交类产品中，甚至会实时转换为几秒钟前、几分钟前、几小时前等不同的格式，这样比直接转换为年、月、日、时、分、秒更友好。本示例就来实现这样一个自定义指令 v-time，将表达式传入的时间戳实时转换为相对时间。

便于演示效果，我们初始化时定义了两个时间。

index.html：

```html
<!DOCTYPE html>
<html>
<head>
    <meta charset="utf-8">
    <title>时间转换指令</title>
</head>
<body>
    <div id="app" v-cloak>
        <div v-time="timeNow"></div>
        <div v-time="timeBefore"></div>
    </div>
    <script src="https://unpkg.com/vue/dist/vue.min.js"></script>
    <script src="time.js"></script>
    <script src="index.js"></script>
</body>
</html>
```

index.js：

```js
var app = new Vue({
    el: '#app',
    data: {
        timeNow: (new Date()).getTime(),
        timeBefore: 1488930695721
    }
});
```

timeNow 是目前的时间，timeBefore 是一个写死的时间：2017-03-08。

提示　　本示例所用的时间戳都是毫秒级的，如服务端返回秒级时间戳需要乘以 1000 后再使用。

分析一下时间转换的逻辑：

- 1 分钟以前，显示"刚刚"。
- 1 分钟~1 小时之间，显示"xx 分钟前"。
- 1 小时~1 天之间，显示"xx 小时前"。
- 1 天~1 个月（31 天）之间，显示"xx 天前"。
- 大于 1 个月，显示"xx 年 xx 月 xx 日"。

为了使判断逻辑更简单，统一使用时间戳进行大小判断。在写指令 v-time 之前，需要先写一系列与时间相关的函数，我们声明一个对象 Time，把它们都封装在里面。

time.js：

```
var Time = {
    // 获取当前时间戳
    getUnix: function () {
        var date = new Date();
        return date.getTime();
    },
    // 获取今天 0 点 0 分 0 秒的时间戳
    getTodayUnix: function () {
        var date = new Date();
        date.setHours(0);
        date.setMinutes(0);
        date.setSeconds(0);
        date.setMilliseconds(0);
        return date.getTime();
    },
    // 获取今年 1 月 1 日 0 点 0 分 0 秒的时间戳
    getYearUnix: function () {
        var date = new Date();
        date.setMonth(0);
        date.setDate(1);
        date.setHours(0);
        date.setMinutes(0);
        date.setSeconds(0);
        date.setMilliseconds(0);
        return date.getTime();
    },
    // 获取标准年月日
    getLastDate: function(time) {
        var date = new Date(time);
        var month = date.getMonth() + 1 < 10 ? '0' + (date.getMonth() + 1) : date.getMonth() + 1;
        var day = date.getDate() < 10 ? '0' + date.getDate() : date.getDate();
        return date.getFullYear() + '-' + month + "-" + day;
    },
    // 转换时间
    getFormatTime: function(timestamp) {
        var now = this.getUnix();                //当前时间戳
        var today = this.getTodayUnix();         //今天 0 点时间戳
        var year = this.getYearUnix();           //今年 0 点时间戳
```

```
        var timer = (now - timestamp) / 1000;    //转换为秒级时间戳
        var tip = '';

        if (timer <= 0) {
            tip = '刚刚';
        } else if (Math.floor(timer/60) <= 0) {
            tip = '刚刚';
        } else if (timer < 3600) {
            tip = Math.floor(timer/60) + '分钟前';
        } else if (timer >= 3600 && (timestamp - today >= 0) ) {
            tip = Math.floor(timer/3600) + '小时前';
        } else if (timer/86400 <= 31) {
            tip = Math.ceil(timer/86400) + '天前';
        } else {
            tip = this.getLastDate(timestamp);
        }
        return tip;
    }
};
```

提 示

如果你对 JavaScript 的 Date 类型不了解,可以到 MDN 查阅学习 Date 常用的 API:
https://developer.mozilla.org/zh-CN/docs/Web/JavaScript/Reference/Global_Objects/Date

Time.getFormatTime()方法就是自定义指令 v-time 所需要的,入参为毫秒级时间戳,返回已经整理好时间格式的字符串。

最后在 time.js 里补全剩余的代码:

```
Vue.directive('time', {
    bind: function (el, binding) {
        el.innerHTML = Time.getFormatTime(binding.value);
        el.__timeout__ = setInterval(function () {
            el.innerHTML = Time.getFormatTime(binding.value);
        }, 60000);
    },
    unbind: function (el) {
        clearInterval(el.__timeout__);
        delete el.__timeout__;
    }
});
```

在 bind 钩子里,将指令 v-time 表达式的值 binding.value 作为参数传入 Time.getFormatTime()方法得到格式化时间,再通过 el.innerHTML 写入指令所在元素。定时器 el.__timeout__ 每分钟触发一次,更新时间,并且在 unbind 钩子里清除掉。

总结：在编写自定义指令时，给 DOM 绑定一次性事件等初始动作，建议在 bind 钩子内完成，同时要在 unbind 内解除相关绑定。在自定义指令里，理论上可以任意操作 DOM，但这又违背 Vue.js 的初衷，所以对于大幅度的 DOM 变动，应该使用组件。

练习 1：开发一个自定义指令 v-birthday，接收一个出生日期的时间戳，将它转换为已经出生了 xxx 天。

练习 2：扩展练习 1 的自定义指令 v-birthday，将出生了 xxx 天转换为具体年龄，比如 25 岁 8 个月 10 天。

第 2 篇　进阶篇

基础篇的章节内容基本涵盖了 Vue.js 2.x 最常用的功能。如果不需要前端路由和自动化工程，那么你已经可以利用这些内容做一些中小型项目了。

从下一章开始，介绍的内容会由浅入深，逐步向前端工程化迈进，使用到的知识点也逐渐增加，比如 NPM、Node.js、ES2015。当然，你完全不用担心，所有知识点都会详细讲解到。

如果你是编程新手、前端入门者，或者刚从后端转到前端，那建议你在阅读后面章节前先巩固一下基础篇所讲的知识点，尤其是组件的章节，最好是先练习一些典型的题目，加深对 Vue.js 基础知识的理解。以下是推荐的一个小项目。

项目：调查问卷 WebApp。

描述：制作一个简单的调查问卷 HTML 5 小应用，每页有一道题目，题目可以是单选题、多选题、填写题等。最终效果如下图所示。

说明：每一页可以通过 v-show 或 v-if 在切换步骤时显示，点击重置，当前页的控件还原到初始状态。要对每页的数据进行校验，比如单选题必须要选择，多选题最少选择 2 项，最多选择 3 项，文本框输入不能少于 100 字，若当前页不满足验证要求，则下一步的按钮置灰，不可点击。

要求：按钮要制作成组件，可以控制颜色、状态（禁用），点击后传递一个自定义事件 on-click。

如果你可以轻松完成这个小练习，或者已经迫不及待地想阅读下一章节，那就做好准备，来探索新的内容吧！

第 9 章

Render 函数

Vue.js 2.x 与 Vue.js 1.x 最大的区别就在于 2.x 使用了 Virtual Dom（虚拟 DOM）来更新 DOM 节点，提升渲染性能。

虽然前面章节我们的组件模板都是写在 template 选项里的，但是在 Vue.js 编译时，都会解析为 Virtual Dom。

本章我们就来探索 Vue.js 用于实现 Virtual Dom 的 Render 函数用法，在介绍 Render 函数前，我们先来看看什么是 Virtual Dom。

9.1 什么是 Virtual Dom

React 和 Vue 2 都使用了 Virtual Dom 技术，Virtual Dom 并不是真正意义上的 DOM，而是一个轻量级的 JavaScript 对象，在状态发生变化时，Virtual Dom 会进行 Diff 运算，来更新只需要被替换的 DOM，而不是全部重绘。

与 DOM 操作相比，Virtual Dom 是基于 JavaScript 计算的，所以开销会小很多。图 9-1 演示了 Virtual Dom 运行的过程。

正常的 DOM 节点在 HTML 中是这样的：

```
<div id="main">
    <p>文本内容</p>
    <p>文本内容</p>
</div>
```

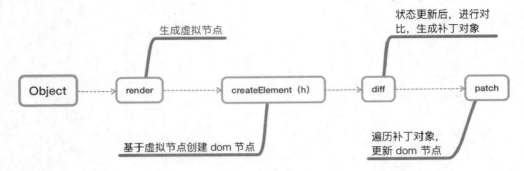

图 9-1 Virtual Dom 运行过程

用 Virtual Dom 创建的 JavaScript 对象一般会是这样的：

```
var vNode = {
    tag: 'div',
    attributes: {
        id: 'main'
    },
    children: [
        // p 节点
    ]
}
```

vNode 对象通过一些特定的选项描述了真实的 DOM 结构。

在 Vue.js 2 中，Virtual Dom 就是通过一种 VNode 类表达的，每个 DOM 元素或组件都对应一个 VNode 对象，在 Vue.js 源码中是这样定义的：

```
export interface VNode {
    tag?: string;
    data?: VNodeData;
    children?: VNode[];
    text?: string;
    elm?: Node;
    ns?: string;
    context?: Vue;
    key?: string | number;
    componentOptions?: VNodeComponentOptions;
    componentInstance?: Vue;
    parent?: VNode;
    raw?: boolean;
    isStatic?: boolean;
    isRootInsert: boolean;
    isComment: boolean;
}
```

具体含义如下：

- tag   当前节点的标签名。
- data   当前节点的数据对象。

VNodeData 代码如下：

```
export interface VNodeData {
    key?: string | number;
    slot?: string;
    scopedSlots?: { [key: string]: ScopedSlot };
    ref?: string;
    tag?: string;
    staticClass?: string;
    class?: any;
    staticStyle?: { [key: string]: any };
    style?: Object[] | Object;
    props?: { [key: string]: any };
    attrs?: { [key: string]: any };
    domProps?: { [key: string]: any };
    hook?: { [key: string]: Function };
    on?: { [key: string]: Function | Function[] };
    nativeOn?: { [key: string]: Function | Function[] };
    transition?: Object;
    show?: boolean;
    inlineTemplate?: {
      render: Function;
      staticRenderFns: Function[];
    };
    directives?: VNodeDirective[];
    keepAlive?: boolean;
}
```

- children   子节点，数组，也是 VNode 类型。
- text   当前节点的文本，一般文本节点或注释节点会有该属性。
- elm   当前虚拟节点对应的真实 DOM 节点。
- ns   节点的 namespace。
- context   编译作用域。。
- functionalContext   函数化组件的作用域。
- key   节点的 key 属性，用于作为节点的标识，有利于 patch 的优化。
- componentOptions   创建组件实例时会用到的选项信息。
- child   当前节点对应的组件实例。
- parent   组件的占位节点。
- raw   原始 html。

- isStatic　静态节点的标识。
- isRootInsert　是否作为根节点插入，被<transition>包裹的节点，该属性的值为 false。
- isComment　当前节点是否是注释节点。
- isCloned　当前节点是否为克隆节点。
- isOnce　当前节点是否有 v-once 指令。

VNode 主要可以分为如下几类，如图 9-2 所示。

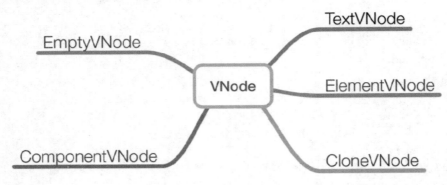

图 9-2　VNode 主要类型

- TextVNode　文本节点。
- ElementVNode　普通元素节点。
- ComponentVNode　组件节点。
- EmptyVNode　没有内容的注释节点。
- CloneVNode　克隆节点，可以是以上任意类型的节点，唯一的区别在于 isCloned 属性为 true。

使用 Virtual Dom 就可以完全发挥 JavaScript 的编程能力。在多数场景中，我们使用 template 就足够了，但在一些特定的场景下，使用 Virtual Dom 会更简单，下节就来介绍 Vue 的 Render 函数的用法。

9.2　什么是 Render 函数

先来看这样一个场景：在很多文章类型的网站中（比如文档、博客）都有区分一级标题、二级标题、三级标题……为方便分享 url，它们都做成了锚点，点击一下，会将内容加在网址后面，以 "#" 分割，如图 9-3 所示。

图中的 "特性" 是一个 <h2> 标签，内容含有一个 <a href="#特性">#</a> 的链接，点击后，url 就带有了锚点信息，别人打开时，会直接聚焦到 "特性" 所在的位置。

第 9 章 Render 函数

图 9-3　带有锚点的标题

如果把它封装为一个组件，一般写法可能会是这样：

```
<div id="app">
    <anchor :level="2" title="特性">特性</anchor>
</div>

<script type="text/x-template" id="anchor">
    <div>
        <h1 v-if="level === 1">
            <a :href="'#' + title">
                <slot></slot>
            </a>
        </h1>
        <h2 v-if="level === 2">
            <a :href="'#' + title">
                <slot></slot>
            </a>
        </h2>
        <h3 v-if="level === 3">
            <a :href="'#' + title">
                <slot></slot>
            </a>
        </h3>
        <h4 v-if="level === 4">
            <a :href="'#' + title">
                <slot></slot>
            </a>
```

```
            </h4>
            <h5 v-if="level === 5">
                <a :href="'#' + title">
                    <slot></slot>
                </a>
            </h5>
            <h6 v-if="level === 6">
                <a :href="'#' + title">
                    <slot></slot>
                </a>
            </h6>
        </div>
    </script>
</div>
<script>
    Vue.component('anchor', {
        template: '#anchor',
        props: {
            level: {
                type: Number,
                required: true
            },
            title: {
                type: String,
                default: ''
            }
        }
    });

    var app = new Vue({
        el: '#app'
    })
</script>
```

这样写没有任何错误，只是缺点很明显：代码冗长，组件的 template 大部分代码是重复的，只是 heading 元素的级别不同，再者必须插入一个根元素<div>，这是组件的要求。

template 写法在大多时候是很好用的，但到了这里使用起来就很别扭。事实上，prop：level 已经具备了 heading 级别的含义，我们更希望能像拼接字符串的形式来构造 heading 元素，比如"h" + this.level。在 Render 函数中的确可以这样做。

下面是使用 Render 函数改写后的代码：

```
<div id="app">
    <anchor :level="2" title="特性">特性</anchor>
```

```
</div>
<script>
    Vue.component('anchor', {
        props: {
            level: {
                type: Number,
                required: true
            },
            title: {
                type: String,
                default: ''
            }
        },
        render: function (createElement) {
            return createElement(
                'h' + this.level,
                [
                    createElement(
                        'a',
                        {
                            domProps: {
                                href: '#' + this.title
                            }
                        },
                        this.$slots.default
                    )
                ]
            )
        }
    });

    var app = new Vue({
        el: '#app'
    })
</script>
```

Render 函数通过 createElement 参数来创建 Virtual Dom，结构精简了很多。在第 7 章组件中介绍 slot 时，有提到过访问 slot 的用法，使用场景就是在 Render 函数。

Render 函数所有神奇的地方都在这个 createElement 里，下一节我们就来介绍它的详细配置和用法。

9.3　createElement 用法

9.3.1　基本参数

createElement 构成了 Vue Virtual Dom 的模板，它有 3 个参数：

```
createElement(
    // {String | Object | Function}
    // 一个 HTML 标签，组件选项，或一个函数
    //必须 Return 上述其中一个
    'div',
    // {Object}
    // 一个对应属性的数据对象，可选
    // 您可以在 template 中使用
    {
        // 稍后详细介绍
    },
    // {String | Array}
    // 子节点（VNodes），可选
    [
        createElement('h1', 'hello world'),
        createElement(MyComponent, {
            props: {
                someProp: 'foo'
            }
        }),
        'bar'
    ]
)
```

第一个参数必选，可以是一个 HTML 标签，也可以是一个组件或函数；第二个是可选参数，数据对象，在 template 中使用。第三个是子节点，也是可选参数，用法一致。

对于第二个参数"数据对象"，具体的选项如下：

```
{
    //和 v-bind:class 一样的 API
    'class': {
        foo: true,
        bar: false
    },
    //和 v-bind:style 一样的 API
    style: {
        color: 'red',
```

```
        fontSize: '14px'
},
// 正常的 HTML 特性
attrs: {
    id: 'foo'
},
//组件 props
props: {
    myProp: 'bar'
},
// DOM 属性
domProps: {
    innerHTML: 'baz'
},
// 自定义事件监听器"on"
//不支持如 v-on:keyup.enter 的修饰器
// 需要手动匹配 keyCode
on: {
    click: this.clickHandler
},
// 仅对于组件，用于监听原生事件
// 而不是组件使用 vm.$emit 触发的自定义事件
nativeOn: {
    click: this.nativeClickHandler
},
// 自定义指令
directives: [
    {
        name: 'my-custom-directive',
        value: '2',
        expression: '1 + 1',
        arg: 'foo',
        modifiers: {
            bar: true
        }
    }
],
// 作用域 slot
// { name: props => VNode | Array<VNode> }
scopedSlots: {
    default: props => h('span', props.text)
},
// 如果子组件有定义 slot 的名称
```

```
    slot: 'name-of-slot'
    // 其他特殊顶层属性
    key: 'myKey',
    ref: 'myRef'
}
```

以往在 template 里，我们都是在组件的标签上使用形容 v-bind:class、v-bind:style、v-on:click 这样的指令，在 Render 函数都将其写在了数据对象里，比如下面的组件，使用传统的 template 写法是：

```
<div id="app">
    <ele></ele>
</div>
<script>
    Vue.component('ele', {
        template: '\
            <div id="element" \
            :class="{show: show}" \
            @click="handleClick">文本内容</div>',
        data: function () {
            return {
                show: true
            }
        },
        methods: {
            handleClick: function () {
                console.log('clicked!');
            }
        }
    });

    var app = new Vue({
        el: '#app'
    })
</script>
```

使用 Render 改写后的代码如下：

```
<div id="app">
    <ele></ele>
</div>
<script>
    Vue.component('ele', {
        render: function (createElement) {
```

```
            return createElement(
                'div',
                {
                    // 动态绑定 class,同:class
                    class: {
                        'show': this.show
                    },
                    // 普通 html 特性
                    attrs: {
                        id: 'element'
                    },
                    //给 div 绑定 click 事件
                    on: {
                        click: this.handleClick
                    }
                },
                '文本内容'
            )
        },
        data: function () {
            return {
                show: true
            }
        },
        methods: {
            handleClick: function () {
                console.log('clicked!');
            }
        }
    });

    var app = new Vue({
        el: '#app'
    })
</script>
```

就此例而言，template 的写法明显比 Render 写法要可读而且简洁，所以要在合适的场景使用 Render 函数，否则只会增加负担。

9.3.2 约束

所有的组件树中，如果 VNode 是组件或含有组件的 slot，那么 VNode 必须唯一。所以下面的两个示例都是错误的。

示例一,重复使用组件,代码如下:

```html
<div id="app">
    <ele></ele>
</div>
<script>
    // 局部声明组件
    var Child = {
        render: function(createElement) {
            return createElement('p', 'text');
        }
    };
    Vue.component('ele', {
        render: function (createElement) {
            // 创建一个子节点,使用组件 Child
            var ChildNode = createElement(Child);
            return createElement('div', [
                ChildNode,
                ChildNode
            ]);
        }
    });

    var app = new Vue({
        el: '#app'
    })
</script>
```

示例二,重复使用含有组件的 slot,代码如下:

```html
<div id="app">
    <ele>
        <div>
            <Child></Child>
        </div>
    </ele>
</div>
<script>
    // 全局注册组件
    Vue.component('Child', {
        render: function (createElement) {
            return createElement('p', 'text');
        }
    });
```

```
    Vue.component('ele', {
        render: function (createElement) {
            return createElement('div', [
                this.$slots.default,
                this.$slots.default
            ]);
        }
    });

    var app = new Vue({
        el: '#app'
    })
</script>
```

这两个示例都期望在子节点内渲染出两个 Child 组件,也就是两个<p>text</p> 节点,实际预览时只渲染出了一个,因为在这种情况下,VNode 受到了约束。

对于重复渲染多个组件(或元素)的方法有很多,比如下面的示例:

```
<div id="app">
    <ele></ele>
</div>
<script>
    // 局部声明组件
    var Child = {
        render: function(createElement) {
            return createElement('p', 'text');
        }
    };
    Vue.component('ele', {
        render: function (createElement) {
            return createElement('div',
                Array.apply(null, {
                    length: 5
                }).map(function () {
                    return createElement(Child);
                })
            );
        }
    });

    var app = new Vue({
        el: '#app'
    })
</script>
```

上例通过一个循环和工厂函数就可以渲染 5 个重复的子组件 Child。对于含有组件的 slot，复用就要稍微复杂一点了，需要将 slot 的每个子节点都克隆一份。示例代码如下：

```
<div id="app">
    <ele>
        <div>
            <Child></Child>
        </div>
    </ele>
</div>
<script>
    // 全局注册组件
    Vue.component('Child', {
        render: function (createElement) {
            return createElement('p', 'text');
        }
    });
    Vue.component('ele', {
        render: function (createElement) {
            // 克隆 slot 节点的方法
            function cloneVNode (vnode) {
                // 递归遍历所有子节点，并克隆
                const clonedChildren = vnode.children &&
                vnode.children.map(function(vnode) {
                    return cloneVNode(vnode);
                });
                const cloned = createElement(
                    vnode.tag,
                    vnode.data,
                    clonedChildren
                );
                cloned.text = vnode.text;
                cloned.isComment = vnode.isComment;
                cloned.componentOptions = vnode.componentOptions;
                cloned.elm = vnode.elm;
                cloned.context = vnode.context;
                cloned.ns = vnode.ns;
                cloned.isStatic = vnode.isStatic;
                cloned.key = vnode.key;

                return cloned;
            }
```

```
            const vNodes = this.$slots.default;
            const clonedVNodes = vNodes.map(function(vnode) {
                return cloneVNode(vnode);
            });

            return createElement('div', [
                vNodes,
                clonedVNodes
            ]);
        }
    });

    var app = new Vue({
        el: '#app'
    })
</script>
```

在 Render 函数里创建了一个 cloneVNode 的工厂函数,通过递归将 slot 所有子节点都克隆了一份,并对 VNode 的关键属性也进行复制。

深度克隆 slot 的做法有点偏黑科技,不过在一般业务中几乎不会遇到这样的需求,主要还是运用在独立组件中。

9.3.3 使用 JavaScript 代替模板功能

在 Render 函数中,不再需要 Vue 内置的指令,比如 v-if、v-for,当然,也没办法使用它们。无论要实现什么功能,都可以用原生 JavaScript。比如 v-if 和 v-else 可以这样写:

```
<div id="app">
    <ele :show="show"></ele>
    <button @click="show = !show">切换 show</button>
</div>
<script>
    Vue.component('ele', {
        render: function (createElement) {
            if (this.show) {
                return createElement('p', 'show 的值为 true');
            } else {
                return createElement('p', 'show 的值为 false');
            }
        },
        props: {
            show: {
                type: Boolean,
                default: false
```

```
        }
      }
    });

    var app = new Vue({
        el: '#app',
        data: {
            show: false
        }
    })
</script>
```

上例直接使用了 JavaScript 的 if 和 else 语句来完成逻辑判断。对于 v-for，可以用一个简单的 for 循环来实现：

```
<div id="app">
    <ele :list="list"></ele>
</div>
<script>
    Vue.component('ele', {
        render: function (createElement) {
            var nodes = [];
            for (var i = 0; i < this.list.length; i++) {
                nodes.push(createElement('p', this.list[i]));
            }
            return createElement('div', nodes);
        },
        props: {
            list: {
                type: Array
            }
        }
    });

    var app = new Vue({
        el: '#app',
        data: {
            list: [
                '《Vue.js 实战》',
                '《JavaScript 高级程序设计》',
                '《JavaScript 语言精粹》'
            ]
        }
    })
</script>
```

在一开始接触 Render 写法时，可能会有点不适应，毕竟这种用 createElement 创建 DOM 节点的方法不够直观和可读，而且受 Vue 内置指令的影响，有时会绕不过弯。不过只要把它当作 JavaScript 一个普通的函数来使用，写习惯后就没有那么难理解了，说到底，它只是 JavaScript 而已。比如下面的示例展示了 JavaScript 的 if、else 语句和数组 map 方法充分配合使用来渲染一个列表。示例代码如下：

```html
<div id="app">
    <ele :list="list"></ele>
    <button @click="handleClick">显示列表</button>
</div>
<script>
    Vue.component('ele', {
        render: function (createElement) {
            if (this.list.length) {
                return createElement('ul', this.list.map(function (item) {
                    return createElement('li', item);
                }));
            } else {
                return createElement('p', '列表为空');
            }
        },
        props: {
            list: {
                type: Array,
                default: function () {
                    return [];
                }
            }
        }
    });

    var app = new Vue({
        el: '#app',
        data: {
            list: []
        },
        methods: {
            handleClick: function () {
                this.list = [
                    '《Vue.js 实战》',
                    '《JavaScript 高级程序设计》',
                    '《JavaScript 语言精粹》'
                ];
```

```
            }
        }
    })
</script>
```

首先是判断 prop：list 是否为空，如果是空，就渲染一个"列表为空"的<p>元素；如果不为空数组，那就把每一项作为<li>渲染，放在<ul>下。

提 示

map()方法是快速改变数组结构，返回了一个新数组，如果你不熟悉数组的这种链式操作（map 常和 filter、sort 等方法一起使用，因为它们返回的都是新数组），可以使用简单的 for 循环，这样更容易理解。

上例的 Render 函数对应的 template 写法如下：

```
<ul v-if="list.length">
    <li v-for="item in list">{{ item }}</li>
</ul>
<p v-else>列表为空</p>
```

Render 函数里也没有与 v-model 对应的 API，需要自己来实现逻辑。示例代码如下：

```
<div id="app">
    <ele></ele>
</div>
<script>
    Vue.component('ele', {
        render: function (createElement) {
            var _this = this;
            return createElement('div', [
                createElement('input', {
                    domProps: {
                        value: this.value
                    },
                    on: {
                        input: function (event) {
                            _this.value = event.target.value;
                        }
                    }
                }),
                createElement('p', 'value: ' + this.value)
            ])
        },
        data: function () {
            return {
```

```
            value: ''
        }
    }
});

var app = new Vue({
    el: '#app'
})
</script>
```

事实上，v-model 就是 prop：value 和 event：input 组合使用的一个语法糖，虽然在 Render 里写起来比较复杂，但是可以自由控制，深入到更底层。

上例的 Render 函数对应的 template 写法如下：

```
<div>
    <input v-model="value">
    <p>value: {{ value }}</p>
</div>
```

对于事件修饰符和按键修饰符，基本也需要自己实现，表 9-1 列举了大部分修饰符对应的实现方案。

表 9-1　部分事件修饰符和按键修饰符对应的句柄

修饰符	对应的句柄
.stop	event.stopPropagation()
.prevent	event.preventDefault()
.self	if (event.target !== event.currentTarget) return
.enter、.13	if (event.keyCode !== 13) return 替换 13 位需要的 keyCode
.ctrl、.alt、.shift、.meta	if (!event.ctrlKey) return 根据需要替换 ctrlKey 为 altKey、shiftKey 或 metaKey

对于事件修饰符.capture 和.once，Vue 提供了特殊的前缀，可以直接写在 on 的配置里，如表 9-2 所示。

表 9-2　.capture 和.once 事件修饰符的前缀

修饰符	前缀
.capture	!
.once	~
.capture.once 或.once.capture	~!

写法如下：

```
on: {
    '!click': this.doThisInCapturingMode,
    '~keyup': this.doThisOnce,
```

```
    '~!mouseover': this.doThisOnceInCapturingMode
}
```

例如，下面的示例简单模拟了聊天发送内容的场景。示例代码如下：

```
<div id="app">
    <ele></ele>
</div>
<script>
    Vue.component('ele', {
        render: function (createElement) {
            var _this = this;
            // 渲染聊天内容列表
            if (this.list.length) {
                var listNode = createElement('ul', this.list.map(function(item) {
                    return createElement('li', item);
                }));
            } else {
                var listNode = createElement('p', '暂无聊天内容');
            }
            return createElement('div', [
                listNode,
                createElement('input', {
                    attrs: {
                        placeholder: '输入内容，按回车键发送'
                    },
                    style: {
                        width: '200px'
                    },
                    on: {
                        keyup: function (event) {
                            // 如果不是回车键，不发送数据
                            if (event.keyCode !== 13) return;
                            //添加输入的内容到聊天列表
                            _this.list.push(event.target.value);
                            // 发送后，清空输入框
                            event.target.value = '';
                        }
                    }
                })
            ])
        },
        data: function () {
            return {
```

```
            value: '',
            list: []
        }
    }
});

var app = new Vue({
    el: '#app'
})
</script>
```

对于 slot，我们已经介绍过可以用 this.$slots 来访问，在 Render 函数中会大量使用，不过没有使用 slot 时，会显示一个默认的内容，这部分逻辑需要我们自己实现。示例代码如下：

```
<div id="app">
    <ele></ele>
    <ele>
        <p>slot 的内容</p>
    </ele>
</div>
<script>
    Vue.component('ele', {
        render: function (createElement) {
            if (this.$slots.default === undefined) {
                return createElement('div', '没有使用 slot 时显示的文本');
            } else {
                return createElement('div', this.$slots.default);
            }
        }
    });

    var app = new Vue({
        el: '#app'
    })
</script>
```

this.$slots.default 等于 undefined，就说明父组件中没有定义 slot，这时可以自定义显示的内容。

9.4 函数化组件

Vue.js 提供了一个 functional 的布尔值选项，设置为 true 可以使组件无状态和无实例，也就是没有 data 和 this 上下文。这样用 render 函数返回虚拟节点可以更容易渲染，因为函数化组件只是一个函数，渲染开销要小很多。

使用函数化组件时，Render 函数提供了第二个参数 context 来提供临时上下文。组件需要的 data、props、slots、children、parent 都是通过这个上下文来传递的，比如 this.level 要改写为 context.props.level，this.$slots.default 改写为 context.children。

例如，下面的示例用函数化组件展示了一个根据数据智能选择不同组件的场景：

```html
<div id="app">
    <smart-item :data="data"></smart-item>
    <button @click="change('img')">切换为图片组件</button>
    <button @click="change('video')">切换为视频组件</button>
    <button @click="change('text')">切换为文本组件</button>
</div>
<script>
    // 图片组件选项
    var ImgItem = {
        props: ['data'],
        render: function (createElement) {
            return createElement('div', [
                createElement('p', '图片组件'),
                createElement('img', {
                    attrs: {
                        src: this.data.url
                    }
                })
            ]);
        }
    };
    // 视频组件选项
    var VideoItem = {
        props: ['data'],
        render: function (createElement) {
            return createElement('div', [
                createElement('p', '视频组件'),
                createElement('video', {
                    attrs: {
                        src: this.data.url,
                        controls: 'controls',
                        autoplay: 'autoplay'
                    }
                })
            ]);
        }
    };
    // 纯文本组件选项
```

```javascript
    var TextItem = {
        props: ['data'],
        render: function (createElement) {
            return createElement('div', [
                createElement('p', '纯文本组件'),
                createElement('p', this.data.text)
            ]);
        }
    };
    Vue.component('smart-item', {
        //函数化组件
        functional: true,
        render: function (createElement, context) {
            // 根据传入的数据，智能判断显示哪种组件
            function getComponent () {
                var data = context.props.data;
                // 判断 prop: data 的 type 字段是属于哪种类型的组件
                if (data.type === 'img')   return ImgItem;
                if (data.type === 'video') return VideoItem;
                return TextItem;
            }
            return createElement(
                getComponent(),
                {
                    props: {
                        //把 smart-item 的 prop: data 传给上面智能选择的组件
                        data: context.props.data
                    }
                },
                context.children
            )
        },
        props: {
            data: {
                type: Object,
                required: true
            }
        }
    })
    var app = new Vue({
        el: '#app',
        data: {
            data: {}
```

```
        },
        methods: {
            // 切换不同类型组件的数据
            change: function (type) {
                if (type === 'img') {
                    this.data = {
                        type: 'img',
                        url: 'https://raw.githubusercontent.com/iview/iview/master/assets/logo.png'
                    }
                } else if (type === 'video') {
                    this.data = {
                        type: 'video',
                        url: 'http://vjs.zencdn.net/v/oceans.mp4'
                    }
                } else if (type === 'text') {
                    this.data = {
                        type: 'text',
                        content: '这是一段纯文本'
                    }
                }
            }
        },
        created: function () {
            // 初始化时，默认设置图片组件的数据
            this.change('img');
        }
    })
</script>
```

代码片段比较长，逐步分析一下实现的内容。ImgItem、VideoItem、TextItem 这 3 个对象分别是图片组件、视频组件和纯文本组件的选项，它们都接收一个 prop：data。在函数化组件 smart-item 里，也有 props：data，通过 getComponent 函数来判断其字段 type 的值，选择这条数据适合渲染的组件。通过 createElement 把 getComponent() 返回的对象设置为第一个参数，然后通过第二个参数把 smart-item 的 data 传递到选择的组件里的 prop：data，组件渲染出不同的内容。

根实例 app 中的方法 change 用来生成不同的数据，通过 3 个 button 来切换。

该示例难理解的地方在于 smart-item 和 3 个功能组件都有 prop：data，它们的传递顺序和原理看起来比较含糊。

函数化组件在业务中并不是很常用，而且也有其他类似的方法来实现，比如上例也可以用组件的 is 特性来动态挂载。总结起来，函数化组件主要适用于以下两个场景：

- 程序化地在多个组件中选择一个。
- 在将 children, props, data 传递给子组件之前操作它们。

9.5　JSX

使用 Render 函数最不友好的地方就是在模板比较简单时，写起来也很复杂，而且难以阅读出 DOM 结构，尤其当子节点嵌套较多时，嵌套的 createElement 就像盖楼一样一层层延伸下去。举一个例子，比如使用 template 书写的模板是：

```
<Anchor :level="1">
    <span>一级</span> 标题
</Anchor>
```

使用 createElement 改写后应该是：

```
return createElement('Anchor', {
    props: {
        level: 1
    }
}, [
    createElement('span', '一级'),
    '标题'
]);
```

为了让 Render 函数更好地书写和阅读，Vue.js 提供了插件 babel-plugin-transform-vue-jsx 来支持 JSX 语法。

JSX 是一种看起来像 HTML，但实际是 JavaScript 的语法扩展，它用更接近 DOM 结构的形式来描述一个组件的 UI 和状态信息，最早在 React.js 里大量应用。

比如上面的 Render 用 JSX 改写后的代码是：

```
new Vue({
    el: '#app',
    render (h) {
        return (
            <Anchor level={1}>
                <span>一级</span> 标题
            </Anchor>
        )
    }
})
```

上面的代码无法直接运行，需要在 webpack 里配置插件 babel-plugin-transform-vue-jsx 编译后才可以，后面章节会介绍到 webpack 的用法。

这里的 render 使用了 ES2015 的语法缩写了函数，也会在后面的章节提到。需要注意的是，参数 h 不能省略，否则使用时会触发错误。

使用 createElement 时，常用的配置示例如下：

```
render(createElement) {
    return createElement('div', {
        props: {
            text: 'some text'
        },
        attrs: {
            id: 'myDiv'
        },
        domProps: {
            innerHTML: 'content'
        },
        on: {
            change: this.changeHandler
        },
        nativeOn: {
            click: this.clickHandler
        },
        class: {
            show: true,
            on: false
        },
        style: {
            color: '#fff',
            background: '#f50'
        },
        key: 'key',
        ref: 'element',
        refInFor: true,
        slot: 'slot'
    })
}
```

上面的示例使用 JSX 后等同于下面的代码:

```
render (h) {
    return (
        <div
            id="myDiv"
            domPropsInnerHTML="content"
            onChange={this.changeHandler}
            nativeOnClick={this.clickHandler}
            class={{ show: true, on: false }}
            style={{ color: '#fff', background: '#f50' }}
            key="key"
```

```
            ref="element"
            refInFor
            slot="slot">
        </div>
    )
}
```

JSX 仍然是 JavaScript 而不是 DOM，如果你的团队不是 JSX 强驱动的，建议还是以模板 template 的方式为主，特殊场景（比如锚点标题）使用 Render 的 createElement 辅助完成。

9.6 实战：使用 Render 函数开发可排序的表格组件

表格可以用来展示大量结构化的数据。本节将以 Render 函数为基础，开发一个可以对表格某一列数据进行排序的表格组件。最终效果如图 9-4 所示。

姓名	年龄 ↑↓	出生日期 ↑↓	地址
王小明	18	1999-02-21	北京市朝阳区芍药居
张小刚	25	1992-01-23	北京市海淀区西二旗
李小红	30	1987-11-10	上海市浦东新区世纪大道
周小伟	26	1991-10-10	深圳市南山区深南大道

图 9-4 可排序表格组件效果图

一个标准的表格由<table>、<thead>、<tbody>、<tr>、<th>、<td>等元素组成。

表格组件的所有内容（表头和行数据）由两个 prop 构成：columns 和 data。两者都是数组，columns 用来描述每列的信息，并渲染在表头 <thead> 内，可以指定某一列是否需要排序；data 是每一行的数据，由 columns 决定每一行里各列的顺序。

按照惯例，先初始化各个文件。

index.html：

```
<!DOCTYPE html>
<html>
<head>
    <meta charset="utf-8">
    <title>可排序的表格组件</title>
    <link rel="stylesheet" type="text/css" href="style.css">
</head>
<body>
    <div id="app" v-cloak>
        <v-table></v-table>
```

```html
    </div>
    <script src="https://unpkg.com/vue/dist/vue.min.js"></script>
    <script src="table.js"></script>
    <script src="index.js"></script>
</body>
</html>
```

index.js：

```js
var app = new Vue({
    el: '#app'
});
```

table.js：

```js
Vue.component('vTable', {
    props: {
        columns: {
            type: Array,
            default: function () {
                return [];
            }
        },
        data: {
            type: Array,
            default: function () {
                return [];
            }
        }
    }
});
```

为了让排序后的 columns 和 data 不影响原始数据，给 v-table 组件的 data 选项添加两个对应的数据，组件所有的操作将在这两个数据上完成，不对原始数据做任何处理：

```js
Vue.component('vTable', {
    // ...
    data: function () {
        return {
            currentColumns: [],
            currentData: []
        }
    }
});
```

columns 的每一项是一个对象，其中 title 和 key 字段是必填的，用来标识这列的表头标题，key 是对应 data 中列内容的字段名。sortable 是选填字段，如果值为 true，说明该列需要排序。在 index.js 中构造数据，比如：

```
var app = new Vue({
    el: '#app',
    data: {
        columns: [
            {
                title: '姓名',
                key: 'name'
            },
            {
                title: '年龄',
                key: 'age',
                sortable: true
            }
        ],
        data: [
            {
                name: '王小明',
                age: 18,
                birthday: '1999-02-21',
                address: '北京市朝阳区芍药居'
            }
        ]
    }
});
```

在 index.html 里，把数据传递给组件 v-table：

```
<v-table :data="data" :columns="columns"></v-table>
```

v-table 组件目前的 prop：columns 和 data 的数据已经从父级传递过来了，不过前面介绍过，v-table 不直接使用它们，而是使用 data 选项的 currentColumns 和 currentData。所以在 v-table 初始化时，需要把 columns 和 data 赋值给 currentColumns 和 currentData。在 v-table 的 methods 选项里定义两个方法用来赋值，并在 mounted 钩子内调用：

```
Vue.component('vTable', {
    // ...
    methods: {
        makeColumns: function () {
            this.currentColumns = this.columns.map(function (col, index) {
```

```
                //添加一个字段标识当前列排序的状态，后续使用
                col._sortType = 'normal';
                //添加一个字段标识当前列在数组中的索引，后续使用
                col._index = index;
                return col;
            });
        },
        makeData: function () {
            this.currentData = this.data.map(function (row, index) {
                //添加一个字段标识当前行在数组中的索引，后续使用
                row._index = index;
                return row;
            });
        }
    },
    mounted () {
        // v-table 初始化时调用
        this.makeColumns();
        this.makeData();
    }
});
```

map() 是 JavaScript 数组的一个方法，根据传入的函数重新构造一个新数组。排序分升序（asc）和降序（desc）两种，而且同时只能对一列数据进行排序，与其他列互斥，为了标识当前列的排序状态，在 map 列添加数据时，默认给每列都添加一个 _sortType 的字段，并且赋值为 normal，表示默认排序（也就是不排序）。在排序后，currentData 每项的顺序可能都会发生变化，所以给 currentColumns 和 currentData 的每个数据都添加 _index 字段，代表当前数据在原始数据中的索引。

有了数据，下面就来用 Render 函数构造虚拟 DOM。

表格的最外层是 <table> 元素，里面包含了 <thead> 表头和 <tbody> 表格主体。thead 是一行多列（一个 <tr>、多个 <th>），tbody 是多行多列（多个 <tr>、多个 <td>）。先由外至内构建出大致的 DOM 结构：

```
Vue.component('vTable', {
    // ...
    render: function(h) {
        var ths = [];
        var trs = [];
        return h('table',[
            h('thead', [
                h('tr', ths)
            ]),
```

```
        h('tbody', trs)
      ])
    }
    // ...
});
```

这里的 h 就是 createElement，只是换了个名称。表格主体 trs 是一个二维数组，数据由 currentColumns 和 CurrentData 组成：

```
render: function(h) {
    var _this = this;
    // ...
    var trs = [];
    this.currentData.forEach(function(row) {
        var tds = [];
        _this.currentColumns.forEach(function(cell) {
            tds.push(h('td', row[cell.key]));
        });
        trs.push(h('tr', tds));
    });
    // ...
}
```

先遍历所有的行，然后在每一行内再遍历各列，最终组合出主体内容节点 trs。
表头的节点 ths 要相对复杂一点，因为有排序的功能：

```
render: function(h) {
    var _this = this;
    var ths = [];
    this.currentColumns.forEach(function(col, index) {
        if (col.sortable) {
            ths.push(h('th', [
                h('span', col.title),
                // 升序
                h('a', {
                    class: {
                        on: col._sortType === 'asc'
                    },
                    on: {
                        click: function () {
                            _this.handleSortByAsc(index)
                        }
                    }
                }, '↑'),
                // 降序
```

```
                    h('a', {
                        class: {
                            on: col._sortType === 'desc'
                        },
                        on: {
                            click: function () {
                                _this.handleSortByDesc(index)
                            }
                        }
                    }, '↓')
                ]));
            } else {
                ths.push(h('th', col.title));
            }
        });
        // ...
    }
```

如果 col.sortable 没有定义，或值为 false，就直接把 col.title 渲染出来，否则除了渲染 title，还加了两个<a>元素来实现升序和降序的操作。handleSortByAsc 和 handleSortByDesc 代码如下：

```
Vue.component('vTable', {
    // ...
    methods: {
        handleSortByAsc: function (index) {
            var key = this.currentColumns[index].key;
            this.currentColumns.forEach(function (col) {
                col._sortType = 'normal';
            });
            this.currentColumns[index]._sortType = 'asc';

            this.currentData.sort(function (a, b) {
                return a[key] > b[key] ? 1 : -1;
            });
        },
        handleSortByDesc: function (index) {
            var key = this.currentColumns[index].key;
            this.currentColumns.forEach(function (col) {
                col._sortType = 'normal';
            });
            this.currentColumns[index]._sortType = 'desc';
```

```
            this.currentData.sort(function (a, b) {
                return a[key] < b[key] ? 1 : -1;
            });
        }
    }
    // ...
});
```

两个方法基本类似（读者可尝试将两个方法合并为一个），一个是升序操作，一个是降序操作，目的都是改变 currentColumns 数组每项的顺序。排序使用了 JavaScript 数组的 sort() 方法，这里之所以返回 1 和-1，而不直接返回 a[key] < b[key]，也就是 true 或 false，是因为在部分浏览器（比如 Safari）对 sort() 的处理不同，而 1 和-1 可以做到兼容。排序前，先将所有列的排序状态都重置为 normal，然后设置当前列的排序状态（asc 或 desc），对应到 render 里<a>元素的 class 名称 on，后面会通过 CSS 来高亮显示当前列的排序状态。

当渲染完表格后，父级修改了 data 数据，比如增加或删除，v-table 的 currentData 也应该更新，如果某列已经存在排序状态，更新后应该直接处理一次排序。其代码如下：

```
Vue.component('vTable', {
    // ...
    watch: {
        data: function () {
            this.makeData();
            var sortedColumn = this.currentColumns.filter(function (col) {
                return col._sortType !== 'normal';
            });

            if (sortedColumn.length > 0) {
                if (sortedColumn[0]._sortType === 'asc') {
                    this.handleSortByAsc(sortedColumn[0]._index);
                } else {
                    this.handleSortByDesc(sortedColumn[0]._index);
                }
            }
        }
    }
    // ...
});
```

通过遍历 currentColumns 来找出是否按某一列进行过排序，如果有，就按照当前排序状态对更新后的数据做一次排序操作。

以下是完整的代码。

index.html：

```html
<!DOCTYPE html>
<html>
<head>
    <meta charset="utf-8">
    <title>可排序的表格组件</title>
    <link rel="stylesheet" type="text/css" href="style.css">
</head>
<body>
    <div id="app" v-cloak>
        <v-table :data="data" :columns="columns"></v-table>
        <button @click="handleAddData">添加数据</button>
    </div>
    <script src="https://unpkg.com/vue/dist/vue.min.js"></script>
    <script src="table.js"></script>
    <script src="index.js"></script>
</body>
</html>
```

index.js：

```js
var app = new Vue({
    el: '#app',
    data: {
        columns: [
            {
                title: '姓名',
                key: 'name'
            },
            {
                title: '年龄',
                key: 'age',
                sortable: true
            },
            {
                title: '出生日期',
                key: 'birthday',
                sortable: true
            },
```

```
                {
                    title: '地址',
                    key: 'address'
                }
            ],
            data: [
                {
                    name: '王小明',
                    age: 18,
                    birthday: '1999-02-21',
                    address: '北京市朝阳区芍药居'
                },
                {
                    name: '张小刚',
                    age: 25,
                    birthday: '1992-01-23',
                    address: '北京市海淀区西二旗'
                },
                {
                    name: '李小红',
                    age: 30,
                    birthday: '1987-11-10',
                    address: '上海市浦东新区世纪大道'
                },
                {
                    name: '周小伟',
                    age: 26,
                    birthday: '1991-10-10',
                    address: '深圳市南山区深南大道'
                }
            ]
        },
        methods: {
            handleAddData: function () {
                this.data.push({
                    name: '刘小天',
                    age: 19,
                    birthday: '1998-05-30',
                    address: '北京市东城区东直门'
                });
            }
        }
    });
```

table.js:

```js
Vue.component('vTable', {
    props: {
        columns: {
            type: Array,
            default: function () {
                return [];
            }
        },
        data: {
            type: Array,
            default: function () {
                return [];
            }
        }
    },
    data: function () {
        return {
            currentColumns: [],
            currentData: []
        }
    },
    render: function(h) {
        var _this = this;
        var ths = [];
        this.currentColumns.forEach(function(col, index) {
            if (col.sortable) {
                ths.push(h('th', [
                    h('span', col.title),
                    h('a', {
                        class: {
                            on: col._sortType === 'asc'
                        },
                        on: {
                            click: function () {
                                _this.handleSortByAsc(index)
                            }
                        }
                    }, '↑'),
                    h('a', {
                        class: {
                            on: col._sortType === 'desc'
```

```
                    },
                    on: {
                        click: function () {
                            _this.handleSortByDesc(index)
                        }
                    }
                }, '↓')
            ]));
        } else {
            ths.push(h('th', col.title));
        }
    });

    var trs = [];
    this.currentData.forEach(function(row) {
        var tds = [];
        _this.currentColumns.forEach(function(cell) {
            tds.push(h('td', row[cell.key]));
        });
        trs.push(h('tr', tds));
    });
    return h('table',[
        h('thead', [
            h('tr', ths)
        ]),
        h('tbody', trs)
    ])
},
methods: {
    makeColumns: function () {
        this.currentColumns = this.columns.map(function (col, index) {
            col._sortType = 'normal';
            col._index = index;
            return col;
        });
    },
    makeData: function () {
        this.currentData = this.data.map(function(row, index) {
            row._index = index;
            return row;
        });
    },
    handleSortByAsc: function (index) {
```

```js
            var key = this.currentColumns[index].key;
            this.currentColumns.forEach(function (col) {
                col._sortType = 'normal';
            });
            this.currentColumns[index]._sortType = 'asc';

            this.currentData.sort(function (a, b) {
                return a[key] > b[key] ? 1 : -1;
            });
        },
        handleSortByDesc: function (index) {
            var key = this.currentColumns[index].key;
            this.currentColumns.forEach(function (col) {
                col._sortType = 'normal';
            });
            this.currentColumns[index]._sortType = 'desc';

            this.currentData.sort(function (a, b) {
                return a[key] < b[key] ? 1 : -1;
            });
        }
    },
    watch: {
        data: function () {
            this.makeData();
            var sortedColumn = this.currentColumns.filter(function (col) {
                return col._sortType !== 'normal';
            });

            if (sortedColumn.length > 0) {
                if (sortedColumn[0]._sortType === 'asc') {
                    this.handleSortByAsc(sortedColumn[0]._index);
                } else {
                    this.handleSortByDesc(sortedColumn[0]._index);
                }
            }
        }
    },
    mounted () {
        this.makeColumns();
        this.makeData();
    }
});
```

style.css：

```css
[v-cloak]{
    display: none;
}
table{
    width: 100%;
    margin-bottom: 24px;
    border-collapse: collapse;
    border-spacing: 0;
    empty-cells: show;
    border: 1px solid #e9e9e9;
}
table th{
    background: #f7f7f7;
    color: #5c6b77;
    font-weight: 600;
    white-space: nowrap;
}
table td, table th{
    padding: 8px 16px;
    border: 1px solid #e9e9e9;
    text-align: left;
}
table th a{
    display: inline-block;
    margin: 0 4px;
    cursor: pointer;
}
table th a.on{
    color: #3399ff;
}
table th a:hover{
    color: #3399ff;
}
```

练习 1：查阅资料，了解表格的<colgroup>和<col>元素用法后，给 v-table 的 columns 增加一个可以设置列宽的 width 字段，并实现该功能。

练习 2：将该示例的 render 写法改写为 template 写法，加以对比，总结出两者的差异性，深刻理解其使用场景。

9.7 实战：留言列表

本节将继续使用 Render 函数来完成一个留言列表的小功能，效果如图 9-5 所示。

图 9-5 留言列表效果图

与之前的几个实战案例不同的是，留言列表更偏向于业务，而之前的实战（数字输入框、标签页、表格）都是独立的功能组件。将留言列表用组件树展示，如图 9-6 所示。

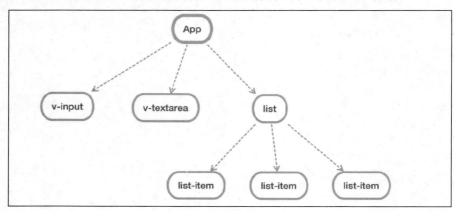

图 9-6 留言列表组件树

先来初始化各个文件。

index.html：

```
<!DOCTYPE html>
<html>
<head>
```

```html
        <meta charset="utf-8">
        <title>留言列表</title>
        <link rel="stylesheet" type="text/css" href="style.css">
    </head>
    <body>
        <div id="app" v-cloak style="width: 500px;margin: 0 auto;">
            <div class="message">

            </div>
        </div>
        <script src="https://unpkg.com/vue/dist/vue.min.js"></script>
        <script src="input.js"></script>
        <script src="list.js"></script>
        <script src="index.js"></script>
    </body>
</html>
```

index.js:

```javascript
var app = new Vue({
    el: '#app'
});
```

input.js:

```javascript
Vue.component('vInput', {

});

Vue.component('vTextarea', {

});
```

发布一条留言，需要的数据有昵称和留言内容，发布操作应该在根实例 app 内完成。留言列表的数据也是从 app 获取的。所以在 index.js 中添加这 3 项数据：

```javascript
var app = new Vue({
    el: '#app',
    data: {
        username: '',
        message: '',
        list: []
    },
    methods: {
        handleSend: function () {
            this.list.push({
```

```
            name: this.username,
            message: this.message
        });
        this.message = '';
    }
  }
});
```

数组 list 存储了所有的留言内容，通过函数 handleSend 给 list 添加一项留言数据，添加成功后，把 textarea 文本框置空。在 index.html 中，使用 v-model 将 username 和 message 双向绑定：

```
<v-input v-model="username"></v-input>
<v-textarea v-model="message"></v-textarea>
```

9.3 节里已经介绍过 Render 函数内的节点如何使用 v-model：动态绑定 value，并且监听 input 事件，把输入的内容通过$emit('input')派发给父组件。所以组件 v-input 的代码如下：

```
Vue.component('vInput', {
    props: {
        value: {
            type: [String, Number],
            default: ''
        }
    },
    render: function (h) {
        var _this = this;
        return h('div', [
            h('span', '昵称：'),
            h('input', {
                attrs: {
                    type: 'text'
                },
                domProps: {
                    value: this.value
                },
                on: {
                    input: function (event) {
                        _this.value = event.target.value;
                        _this.$emit('input', event.target.value);
                    }
                }
            })
        ]);
    }
});
```

v-textarea 与 v-input 基本一致，可查看后面的完整代码。

列表的节点树如图 9-7 所示。

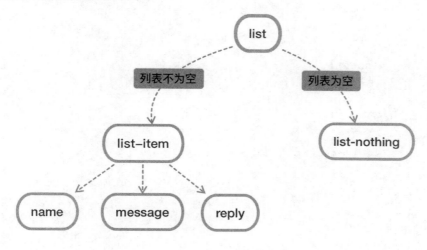

图 9-7　列表树示意图

列表数据 list 为空时，渲染一个"列表为空"的信息提示节点；不为空时，每个 list-item 应包含昵称、留言内容和回复按钮 3 个子节点。list.js 的 render 内容如下：

```
render: function (h) {
    var _this = this;
    var list = [];
    this.list.forEach(function (msg, index) {
        var node = h('div', {
            attrs: {
                class: 'list-item'
            }
        }, [
            h('span', msg.name + ': '),
            h('div', {
                attrs: {
                    class: 'list-msg'
                }
            }, [
                h('p', msg.message),
                h('a', {
                    attrs: {
                        class: 'list-reply'
                    },
                    on: {
                        click: function () {
                            _this.handleReply(index);
```

```
                    }
                }
            }, '回复')
        ])
    ])
    list.push(node);
});
if (this.list.length) {
    return h('div', {
        attrs: {
            class: 'list'
        },
    }, list);
} else {
    return h('div', {
        attrs: {
            class: 'list-nothing'
        }
    }, '留言列表为空');
}
}
```

this.list.forEach 相当于 template 里的 v-for 指令，遍历出每条留言。句柄 handleReply 直接向父组件派发一个事件 reply，父组件（app）接收后，将当前 list-item 的昵称提取，并设置到 v-textarea 内。相关代码如下：

```
// list.js
handleReply: function (index) {
    this.$emit('reply', index);
}

// index.html
<list :list="list" @reply="handleReply"></list>

// index.js
handleReply: function (index) {
    var name = this.list[index].name;
    this.message = '回复@' + name + ': ';
}
```

还有剩余的几个小细节，比如点击回复按钮后，文本框立刻聚焦；提交留言前，做非空判断，读者可在完整代码中继续探索。完整代码如下。

index.html：

```html
<!DOCTYPE html>
<html>
<head>
    <meta charset="utf-8">
    <title>留言列表</title>
    <link rel="stylesheet" type="text/css" href="style.css">
</head>
<body>
    <div id="app" v-cloak style="width: 500px;margin: 0 auto;">
        <div class="message">
            <v-input v-model="username"></v-input>
            <v-textarea v-model="message" ref="message"></v-textarea>
            <button @click="handleSend">发布</button>
        </div>
        <list :list="list" @reply="handleReply"></list>
    </div>
    <script src="https://unpkg.com/vue/dist/vue.min.js"></script>
    <script src="input.js"></script>
    <script src="list.js"></script>
    <script src="index.js"></script>
</body>
</html>
```

index.js：

```js
var app = new Vue({
    el: '#app',
    data: {
        username: '',
        message: '',
        list: []
    },
    methods: {
        handleSend: function () {
            if (this.username === '') {
                window.alert('请输入昵称');
                return;
            }
            if (this.message === '') {
                window.alert('请输入留言内容');
                return;
            }
```

```javascript
            this.list.push({
                name: this.username,
                message: this.message
            });
            this.message = '';
        },
        handleReply: function (index) {
            var name = this.list[index].name;
            this.message = '回复@' + name + ': ';
            this.$refs.message.focus();
        }
    }
});
```

input.js:

```javascript
Vue.component('vInput', {
    props: {
        value: {
            type: [String, Number],
            default: ''
        }
    },
    render: function (h) {
        var _this = this;
        return h('div', [
            h('span', '昵称: '),
            h('input', {
                attrs: {
                    type: 'text'
                },
                domProps: {
                    value: this.value
                },
                on: {
                    input: function (event) {
                        _this.value = event.target.value;
                        _this.$emit('input', event.target.value);
                    }
                }
            })
        ]);
    }
});
```

```js
Vue.component('vTextarea', {
    props: {
        value: {
            type: String,
            default: ''
        }
    },
    render: function (h) {
        var _this = this;
        return h('div', [
            h('span', '留言内容: '),
            h('textarea', {
                attrs: {
                    placeholder: '请输入留言内容'
                },
                domProps: {
                    value: this.value
                },
                ref: 'message',
                on: {
                    input: function (event) {
                        _this.value = event.target.value;
                        _this.$emit('input', event.target.value);
                    }
                }
            })
        ]);
    },
    methods: {
        focus: function () {
            this.$refs.message.focus();
        }
    }
});
```

list.js:

```js
Vue.component('list', {
    props: {
        list: {
            type: Array,
            default: function () {
                return [];
```

```
            }
        }
    },
    render: function (h) {
        var _this = this;
        var list = [];
        this.list.forEach(function (msg, index) {
            var node = h('div', {
                attrs: {
                    class: 'list-item'
                }
            }, [
                h('span', msg.name + ': '),
                h('div', {
                    attrs: {
                        class: 'list-msg'
                    }
                }, [
                    h('p', msg.message),
                    h('a', {
                        attrs: {
                            class: 'list-reply'
                        },
                        on: {
                            click: function () {
                                _this.handleReply(index);
                            }
                        }
                    }, '回复')
                ])
            ])
            list.push(node);
        });
        if (this.list.length) {
            return h('div', {
                attrs: {
                    class: 'list'
                },
            }, list);
        } else {
            return h('div', {
                attrs: {
                    class: 'list-nothing'
```

```
            }
        }, '留言列表为空');
        }
    },
    methods: {
        handleReply: function (index) {
            this.$emit('reply', index);
        }
    }
});
```

style.css:

```css
[v-cloak]{
    display: none;
}
*{
    padding: 0;
    margin: 0;
}
.message{
    width: 450px;
    text-align: right;
}
.message div{
    margin-bottom: 12px;
}
.message span{
    display: inline-block;
    width: 100px;
    vertical-align: top;
}
.message input, .message textarea{
    width: 300px;
    height: 32px;
    padding: 0 6px;
    color: #657180;
    border: 1px solid #d7dde4;
    border-radius: 4px;
    cursor: text;
    outline: none;
}
.message input:focus, .message textarea:focus{
    border: 1px solid #3399ff;
```

```css
    }
    .message textarea{
        height: 60px;
        padding: 4px 6px;
    }
    .message button{
        display: inline-block;
        padding: 6px 15px;
        border: 1px solid #39f;
        border-radius: 4px;
        color: #fff;
        background-color: #39f;
        cursor: pointer;
        outline: none;
    }
    .list{
        margin-top: 50px;
    }
    .list-item{
        padding: 10px;
        border-bottom: 1px solid #e3e8ee;
        overflow: hidden;
    }
    .list-item span{
        display: block;
        width: 60px;
        float: left;
        color: #39f;
    }
    .list-msg{
        display: block;
        margin-left: 60px;
        text-align: justify;
    }
    .list-msg a{
        color: #9ea7b4;
        cursor: pointer;
        float: right;
    }
    .list-msg a:hover{
        color: #39f;
    }
```

```
.list-nothing{
    text-align: center;
    color: #9ea7b4;
    padding: 20px;
}
```

练习1：给每条留言都增加一个删除的功能。

练习2：将该示例的 render 写法改写为 template 写法，加以对比，总结出两者的差异性，深刻理解其使用场景。

9.8 总　　结

本章两个实战的练习题中都有用 template 写法还原 render 函数，目的是充分理解 render 函数的使用场景。如果你已经做了还原，应该会发现使用 template 写法更简单、可读，尤其是第二个示例。的确，这两个实战示例都更适合用 template 来实现，在业务中，生产效率是第一位，所以绝大部分业务代码都应当用 template 来完成。你不用在意性能问题，如果使用了 webpack 做编译（后面章节会介绍），template 都会被预编译为 render 函数。

在本书一开始介绍 Vue.js 时，就提到过它是一个渐进式 JavaScript 框架。Vue.js 的基本用法到本章就结束了，到目前为止，所有的示例都是通过<script>引入 Vue.js 和其他文件来运行的，从下一章开始，将陆续介绍前端工程化和 Vue 生态。

第 10 章

使用 webpack

高效的开发离不开基础工程的搭建。本章主要介绍目前热门的 JavaScript 应用程序的模块打包工具 webpack。在开始学习本章前，需要先安装 Node.js 和 NPM，如果你不熟悉它们，可以先查阅相关资料，完成安装并了解 NPM 最基本的用法。

 本章所介绍的 webpack 是指 webpack 2 版本。

10.1 前端工程化与 webpack

近几年来，前端领域发展迅速，前端的工作早已不再是切几张图那么简单，项目比较大时，可能会多人协同开发。模块化、组件化、CSS 预编译等概念也成了经常讨论的话题。

通常，前端自动化（半自动化）工程主要解决以下问题：

- JavaScript、CSS 代码的合并和压缩。
- CSS 预处理：Less、Sass、Stylus 的编译。
- 生成雪碧图（CSS Sprite）。
- ES 6 转 ES 5。
- 模块化。

……

如果使用过 Gulp，并且了解 RequireJS，上面几个问题应该难不倒你。只需配置几行代码，就可以实现对 JS 代码的合并与压缩。不过，经过 Gulp 合并压缩后的代码仍然是你写的代码，只是局部变量名被替换，一些语法做了转换而已，整体内容并没有发生变化。而本章要介绍的前端工程化工具 webpack，打包后的代码已经不只是你写的代码，其中夹杂了很多 webpack 自身的模块处理代码。因此，学习 webpack 最难的是理解"编译"的这个概念，否则会一直存在一个疑问：为什么要这样做？

图 10-1 是来自 webpack 官方网站（https://webpack.js.org/）经典的模块化示意图。

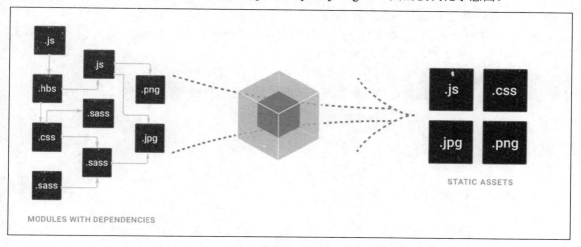

图 10-1　webpack 模块化示意图

左边是在业务中写的各种格式的文件，比如 typescript、less、jpg，还有本章后面要介绍的.vue 格式的文件。这些格式的文件通过特定的加载器（Loader）编译后，最终统一生成为.js、.css、.png 等静态资源文件。在 webpack 的世界里，一张图片、一个 css 甚至一个字体，都称为模块（Module），彼此存在依赖关系，webpack 就是来处理模块间的依赖关系的，并把它们进行打包。

举一个简单的例子，平时加载 CSS 大多通过<link>标签引入 CSS 文件，而在 webpack 里，直接在一个.js 文件中导入，比如：

```
import 'src/styles/index.css';
```

import 是 ES 2015 的语法，这里也可以写成 require('src/styles/index.css') 。在打包时，index.css 会被打包进一个 js 文件里，通过动态创建<style>的形式来加载 css 样式，当然也可以进一步配置，在打包编译时把所有的 css 都提取出来，生成一个 css 的文件，后面会详细介绍。

webpack 的主要适用场景是单页面富应用（SPA）。SPA 通常是由一个 html 文件和一堆按需加载的 js 组成，它的 html 结构可能会非常简单，比如：

```
<!DOCTYPE html>
<html lang="zh-CN">
<head>
    <meta charset="UTF-8">
    <title>webpack app</title>
    <link rel="stylesheet" href="dist/main.css">
</head>
<body>
```

```html
    <div id="app"></div>
    <script type="text/javascript" src="dist/main.js"></script>
</body>
</html>
```

看起来很简单是吧？只有一个<div>节点，所有的代码都集成在了神奇的 main.js 文件中，理论上它可以实现像知乎、淘宝这样大型的项目。

在开始讲解 webpack 的用法前，先介绍两个 ES6 中的语法 export 和 import，因为在后面会大量使用，如果对它不了解，可能会感到很困惑。

export 和 import 是用来导出和导入模块的。一个模块就是一个 js 文件，它拥有独立的作用域，里面定义的变量外部是无法获取的。比如将一个配置文件作为模块导出，示例代码如下：

```js
// config.js
var Config = {
    version: '1.0.0'
};
export { Config };
```

或：

```js
// config.js
export var Config = {
    version: '1.0.0'
};
```

其他类型（比如函数、数组、常量等）也可以导出，比如导出一个函数：

```js
// add.js
export function add(a, b) {
    return a + b;
};
```

模块导出后，在需要使用模块的文件使用 import 再导入，就可以在这个文件内使用这些模块了。示例代码如下：

```js
// main.js
import { Config } from './config.js';
import { add } from './add.js';

console.log(Config);        // { version: '1.0.0' }
console.log(add(1, 1));     // 2
```

以上几个示例中，导入的模块名称都是在 export 的文件中设置的，也就是说用户必须预先知道这个名称叫什么，比如 Config、add。而有的时候，用户不想去了解名称是什么，只是把模块的功能拿来使用，或者想自定义名称，这时可以使用 export default 来输出默认的模块。示例代码如下：

```
// config.js
export default {
    version: '1.0.0'
};

// add.js
export default function (a, b) {
    return a + b;
};

// main.js
import conf from './config.js';
import Add from './add.js';

console.log(conf);       // { version: '1.0.0' }
console.log(Add(1, 1));  // 2
```

如果使用 npm 安装了一些库，在 webpack 中可以直接导入，示例代码如下：

```
import Vue from 'vue';
import $ from 'jquery';
```

上例分别导入了 Vue 和 jQuery 的库，并且命名为 Vue 和$，在这个文件中就可以使用这两个模块。

export 和 import 还有其他的用法，这里不做太详细的介绍，如果有兴趣，可以查阅相关资料进一步学习。

10.2　webpack 基础配置

10.2.1　安装 webpack 与 webpack-dev-server

本节将从基础的 webpack 安装开始介绍，逐步完成对 Vue 工程的配置。在开始学习本节前，先确保已经安装了最新版的 Node.js 和 NPM，并已经了解 NPM 的基本用法。

首先，创建一个目录，比如 demo，使用 NPM 初始化配置：

```
npm init
```

执行后，会有一系列选项，可以按回车键快速确认，完成后会在 demo 目录生成一个 package.json 的文件。

之后在本地局部安装 webpack：

```
npm install webpack --save-dev
```

--save-dev 会作为开发依赖来安装 webpack。安装成功后，在 package.json 中会多一项配置：

```
"devDependencies": {
    "webpack": "^2.3.2"
}
```

接着需要安装 webpack-dev-server，它可以在开发环境中提供很多服务，比如启动一个服务器、热更新、接口代理等，配置起来也很简单。同样，在本地局部安装：

```
npm install webpack-dev-server --save-dev
```

安装完成后，最终的 package.json 文件内容为：

```
{
  "name": "demo",
  "version": "1.0.0",
  "description": "",
  "main": "index.js",
  "scripts": {
    "test": "echo \"Error: no test specified\" && exit 1"
  },
  "author": "",
  "license": "ISC",
  "devDependencies": {
    "webpack": "^2.3.2",
    "webpack-dev-server": "^2.4.2"
  }
}
```

如果你的 devDependencies 中包含 webpack 和 webpack-dev-server，恭喜你，已经安装成功，很快就可以启动 webpack 工程了。

10.2.2 就是一个 js 文件而已

接下来需要了解 webpack 的一些核心概念。

归根到底，webpack 就是一个 .js 配置文件，你的架构好或差都体现在这个配置里，随着需求的不断出现，工程配置也是逐渐完善的。我们从浅入深，一步步来支持更多的功能。

首先在目录 DEMO 下创建一个 js 文件：webpack.config.js，并初始化它的内容：

```
var config = {

};

module.exports = config;
```

这里的 module.exports = config; 相当于 export default config;。由于目前还没有安装支持 ES6 的编译插件，因此不能直接使用 ES6 的语法，否则会报错。

第 10 章 使用 webpack

然后在 package.json 的 scripts 里增加一个快速启动 webpack-dev-server 服务的脚本：

```
{
  // ...
  "scripts": {
    "test": "echo \"Error: no test specified\" && exit 1",
    "dev": "webpack-dev-server --open --config webpack.config.js"
  },
  // ...
}
```

当运行 npm run dev 命令时，就会执行 webpack-dev-server --open --config webpack.config.js 命令。其中 --config 是指向 webpack-dev-server 读取的配置文件路径，这里直接读取我们在上一步创建的 webpack.config.js 文件。--open 会在执行命令时自动在浏览器打开页面，默认地址是 127.0.0.1:8080，不过 IP 和端口都是可以配置的，比如：

```
{
  "scripts": {
    "dev": "webpack-dev-server --host 172.172.172.1 --port 8888 --open --config webpack.config.js"
  }
}
```

这样访问地址就改为了 172.172.172.1:8888。一般在局域网下，需要让其他同事访问时可以这样配置，否则用默认的 127.0.0.1（localhost）就可以了。

webpack 配置中最重要也是必选的两项是入口（Entry）和出口（Output）。入口的作用是告诉 webpack 从哪里开始寻找依赖，并且编译，出口则用来配置编译后的文件存储位置和文件名。

在 demo 目录下新建一个空的 main.js 作为入口的文件，然后在 webpack.config.js 中进行入口和输出的配置：

```
var path = require('path');

var config = {
    entry: {
        main: './main'
    },
    output: {
        path: path.join(__dirname, './dist'),
        publicPath: '/dist/',
        filename: 'main.js'
    }
};

module.exports = config;
```

entry 中的 main 就是我们配置的单入口，webpack 会从 main.js 文件开始工作。output 中 path 选项用来存放打包后文件的输出目录，是必填项。publicPath 指定资源文件引用的目录，如果你的资源存放在 CDN 上，这里可以填 CDN 的网址。filename 用于指定输出文件的名称。因此，这里配置的 output 意为打包后的文件会存储为 demo/dist/main.js 文件，只要在 html 中引入它就可以了。

在 demo 目录下，新建一个 index.html 作为我们 SPA 的入口：

```html
<!DOCTYPE html>
<html>
<head>
    <meta charset="utf-8">
    <title>webpack App</title>
</head>
<body>
    <div id="app">
        Hello World.
    </div>
    <script type="text/javascript" src="/dist/main.js"></script>
</body>
</html>
```

现在在终端执行下面的命令，就会自动在浏览器中打开页面了：

```
npm run dev
```

如果打开的页面跟图 10-2 一致，那么你已经完成整个工程中最重要的一步了。

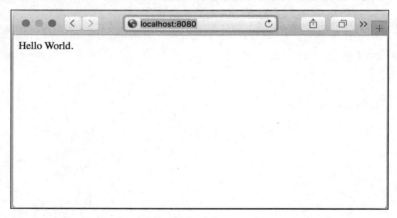

图 10-2　在浏览器中打开 webpack 项目

打开 demo/main.js 文件，添加一行 JavaScript 代码来修改页面的内容：

```
document.getElementById('app').innerHTML = 'Hello webpack.';
```

保存文件，回到刚才打开的页面，发现页面内容已经变为了"Hello webpack."。注意，此时并没有刷新浏览器，就已经自动更新了，这就是 webpack-dev-server 的热更新功能，它通过建立一个 WebSocket 连接来实时响应代码的修改。

在本章第 1 节中介绍过：学习 webpack 最难的是理解它"编译"的这个概念。我们来看一下 webpack 编译出的/dist/main.js 究竟是什么。在 Chrome 浏览器开发者工具的 network 视图中查看 main.js 的内容，如图 10-3 所示。

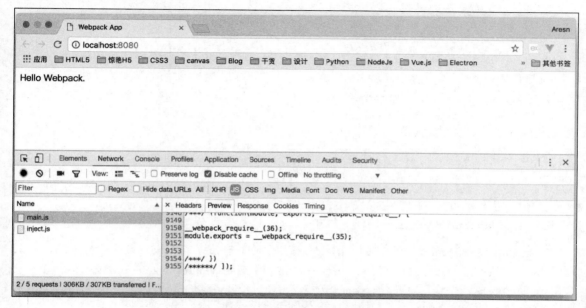

图 10-3　编译后的 main.js 文件

是不是很震惊，我们只写了一行 JS 代码，却编译出了 9000 多行。不过不用担心，这里面很多都是 webpack-dev-server 的功能，只在开发时有效，在生产环境下编译就不会这么臃肿了。比如执行下面的命令进行打包：

```
webpack --progress --hide-modules
```

这时会生成一个 demo/dist/main.js 文件，它只有不到 100 行，而且是没有压缩的。

10.2.3　逐步完善配置文件

10.2.2 小节通过配置入口（Entry）和出口（Output）已经可以启动 webpack 项目了，不过这并不是 webpack 的特点，如果它只有这些作用，根本就不用这么麻烦。本节将对文件 webpack.config.js 进一步配置，来实现更强大的功能。

在 webpack 的世界里，每个文件都是一个模块，比如.css、.js、.html、.jpg、.less 等。对于不同的模块，需要用不同的加载器（Loaders）来处理，而加载器就是 webpack 最重要的功能。通过安装不同的加载器可以对各种后缀名的文件进行处理，比如现在要写一些 CSS 样式，就要用到 style-loader 和 css-loader。下面就通过 NPM 来安装它们：

```
npm install css-loader --save-dev
npm install style-loader --save-dev
```

安装完成后，在 webpack.config.js 文件里配置 Loader，增加对.css 文件的处理：

```
var config = {
    // ...
    module: {
        rules: [
            {
                test: /\.css$/,
                use: [
                    'style-loader',
                    'css-loader'
                ]
            }
        ]
    }
};

module.exports = config;
```

在 module 对象的 rules 属性中可以指定一系列的 loaders，每一个 loader 都必须包含 test 和 use 两个选项。这段配置的意思是说，当 webpack 编译过程中遇到 require()或 import 语句导入一个后缀名为.css 的文件时，先将它通过 css-loader 转换，再通过 style-loader 转换，然后继续打包。use 选项的值可以是数组或字符串，如果是数组，它的编译顺序就是从后往前。

在 demo 目录下新建一个 style.css 的文件，并在 main.js 中导入：

```
/* style.css */
#app{
    font-size: 24px;
    color: #f50;
}

// main.js
import './style.css';
document.getElementById('app').innerHTML = 'Hello webpack.';
```

重新执行 npm run dev 命令，可以看到页面中的文字已经变成红色，并且字号也大了，如图 10-4 所示。

可以看到，CSS 是通过 JavaScript 动态创建<style>标签来写入的，这意味着样式代码都已经编译在了 main.js 文件里，但在实际业务中，可能并不希望这样做，因为项目大了样式会很多，都放在 JS 里太占体积，还不能做缓存。这时就要用到 webpack 最后一个重要的概念——插件（Plugins）。

webpack 的插件功能很强大而且可以定制。这里我们使用一个 extract-text-webpack-plugin 的插件来把散落在各地的 css 提取出来，并生成一个 main.css 的文件，最终在 index.html 里通过<link>的形式加载它。

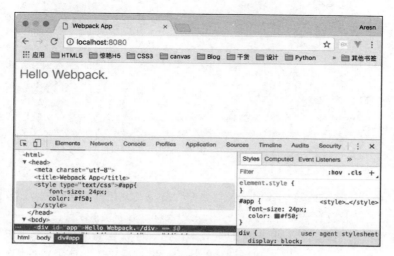

图 10-4　使用 css 后的效果

通过 NPM 安装 extract-text-webpack-plugin 插件：

npm install extract-text-webpack-plugin --save-dev

然后在配置文件中导入插件，并改写 loader 的配置：

```
// 导入插件
var ExtractTextPlugin = require('extract-text-webpack-plugin');

var config = {
    // ...
    module: {
        rules: [
            {
                test: /\.css$/,
                use: ExtractTextPlugin.extract({
                    use: 'css-loader',
                    fallback: 'style-loader'
                })
            }
        ]
    },
    plugins: [
        // 重命名提取后的 css 文件
        new ExtractTextPlugin("main.css")
    ]
};

module.exports = config;
```

```
// index.html
<!DOCTYPE html>
<html>
<head>
    <meta charset="utf-8">
    <title>webpack App</title>
    <link rel="stylesheet" type="text/css" href="/dist/main.css">
</head>
……
```

插件还可以进行丰富的配置,我们会在后面结合 Vue 使用时详细介绍。现在重新启动服务,就可以看到<style> 已经没有了,通过<link> 引入的 main.css 文件已经生效。

webpack 虽然概念比较新,看似复杂,但它只不过是一个 js 配置文件,只要搞清楚入口(Entry)、出口(Output)、加载器(Loaders)和插件(Plugins)这 4 个概念,使用起来就不那么困惑了。

10.3 单文件组件与 vue-loader

回顾一下第 7 章关于组件的内容,我们是如何创建并使用一个组件的。如果你练习过几个示例,肯定会觉得在字符串模板 template 选项里拼写字符串 DOM 非常费劲,尤其是用 "\" 换行。Vue.js 是一个渐进式的 JavaScript 框架,在使用 webpack 构建 Vue 项目时,可以使用一种新的构建模式:.vue 单文件组件。

顾名思义,.vue 单文件组件就是一个后缀名为.vue 的文件,在 webpack 中使用 vue-loader 就可以对.vue 格式的文件进行处理。

一个.vue 文件一般包含 3 部分,即<template>、<script>和<style>,如图 10-5 所示。

图 10-5　.vue 单文件

在 component.vue 文件中,<template></template>之间的代码就是该组件的模板 HTML,<style></style>之间的是 CSS 样式,示例中的<style>标签使用了 scoped 属性,表示当前的 CSS 只在这个组件有效,如果不加,那么 div 的样式会应用到整个项目。<style>还可以结合 CSS 预编译一起使用,比如使用 Less 处理可以写成<style lang="less">。

使用.vue 文件需要先安装 vue-loader、vue-style-loader 等加载器并做配置。因为要使用 ES6 语法,还需要安装 babel 和 babel-loader 等加载器。使用 npm 逐个安装以下依赖:

```
npm install --save vue
npm install --save-dev vue-loader
npm install --save-dev vue-style-loader
npm install --save-dev vue-template-compiler
npm install --save-dev vue-hot-reload-api
npm install --save-dev babel
npm install --save-dev babel-loader
npm install --save-dev babel-core
npm install --save-dev babel-plugin-transform-runtime
npm install --save-dev babel-preset-es2015
npm install --save-dev babel-runtime
```

安装完成后,修改配置文件 webpack.config.js 来支持对.vue 文件及 ES6 的解析:

```
var path = require('path');
var ExtractTextPlugin = require('extract-text-webpack-plugin');

var config = {
    entry: {
        main: './main'
    },
    output: {
        path: path.join(__dirname, './dist'),
        publicPath: '/dist/',
        filename: 'main.js'
    },
    module: {
        rules: [
            {
                test: /\.vue$/,
                loader: 'vue-loader',
                options: {
                    loaders: {
                        css: ExtractTextPlugin.extract({
                            use: 'css-loader',
                            fallback: 'vue-style-loader'
                        })
```

```
                }
            }
        },
        {
            test: /\.js$/,
            loader: 'babel-loader',
            exclude: /node_modules/
        },
        {
            test: /\.css$/,
            use: ExtractTextPlugin.extract({
                use: 'css-loader',
                fallback: 'style-loader'
            })
        }
        ]
    },
    plugins: [
        new ExtractTextPlugin("main.css")
    ]
};

module.exports = config;
```

vue-loader 在编译 .vue 文件时，会对<template>、<script>、<style>分别处理，所以在 vue-loader 选项里多了一项 options 来进一步对不同语言进行配置。比如在对 css 进行处理时，会先通过 css-loader 解析，然后把处理结果再交给 vue-style-loader 处理。当你的技术栈多样化时，可以给<template>、<script>和<style>都指定不同的语言，比如<template lang="jade">和<style lang="less">，然后配置 loaders 就可以了。

在 demo 目录下新建一个名为 .babelrc 的文件，并写入 babel 的配置，webpack 会依赖此配置文件来使用 babel 编译 ES6 代码：

```
{
    "presets": ["es2015"],
    "plugins": ["transform-runtime"],
    "comments": false
}
```

配置好这些后，就可以使用 .vue 文件了。记住，每个 .vue 文件就代表一个组件，组件之间可以相互依赖。

在 demo 目录下新建一个 app.vue 的文件并写入以下内容：

```
<template>
    <div>Hello {{ name }}</div>
```

```
    </template>
    <script>
        export default {
            data () {
                return {
                    name: 'Vue.js'
                }
            }
        }
    </script>
    <style scoped>
        div{
            color: #f60;
            font-size: 24px;
        }
    </style>
```

ES 6 语法提示：

```
        data () { }
```

等同于：

```
        data: function () { }
```

在<template>内写的 HTML 写法完全同 html 文件，不用加 "\" 换行，webpack 最终会把它编译为 Render 函数的形式。写在<style>里的样式，我们已经用插件 extract-text-webpack-plugin 配置过了，最终会统一提取并打包在 main.css 里，因为加了 scoped 属性，这部分样式只会对当前组件 app.vue 有效。

.vue 的组件是没有名称的，在父组件使用时可以对它自定义。写好了组件，就可以在入口 main.js 中使用它了。打开 main.js 文件，把内容替换为下面的代码：

```
// 导入 Vue 框架
import Vue from 'vue';
// 导入 app.vue 组件
import App from './app.vue';

// 创建 Vue 根实例
new Vue({
    el: '#app',
    render: h => h(App)
});
```

ES 6 语法提示：

=>是箭头函数

render: h => h(App)等同于:

```
render: function (h) {
    return h(App)
}
```

也等同于:

```
render: h => {
    return h(App);
}
```

箭头函数里的 this 指向与普通函数是不一样的,箭头函数体内的 this 对象就是定义时所在的对象,而不是使用时所在的对象，比如:

```
function Timer () {
    this.id = 1;

    var _this = this;
    setTimeout(function () {
        console.log(this.id);   // undefined
        console.log(_this.id);  // 1
    }, 1000);

    setTimeout(() => {
        console.log(this.id);   // 1
    }, 2000);
}

var timer = new Timer();
```

执行命令 npm run dev，第一个 Vue 工程就跑起来了。打开 Chrome 调试工具，在 Elements 面板可以看到，<div id="app">已经被组件替换成了:

```
<div data-v-8ecbb1fa>Hello Vue.js</div>
```

对应的 main.css 为:

```
div[data-v-8ecbb1fa]{
    color: #f60;
    font-size: 24px;
}
```

之所以多了一串 data-v-xxx 的内容，是因为使用了<style scoped>功能，如果去掉 scoped，就只剩下<div>Hello Vue.js</div>了。

接下来，在 demo 目录下再新建两个文件，title.vue 和 button.vue。

title.vue：

```vue
<template>
    <h1>
        <a :href="'#' + title">{{ title }}</a>
    </h1>
</template>
<script>
    export default {
        props: {
            title: {
                type: String
            }
        }
    }
</script>
<style scoped>
    h1 a{
        color: #3399ff;
        font-size: 24px;
    }
</style>
```

button.vue：

```vue
<template>
    <button @click="handleClick" :style="styles">
        <slot></slot>
    </button>
</template>
<script>
    export default {
        props: {
            color: {
                type: String,
                default: '#00cc66'
            }
        },
        computed: {
            styles () {
                return {
                    background: this.color
                }
```

```
            }
        },
        methods: {
            handleClick (e) {
                this.$emit('click', e);
            }
        }
    }
</script>
<style scoped>
    button{
        border: 0;
        outline: none;
        color: #fff;
        padding: 4px 8px;
    }
    button:active{
        position: relative;
        top: 1px;
        left: 1px;
    }
</style>
```

改写根实例 app.vue 组件,把 title.vue 和 button.vue 导入进来:

```
<template>
    <div>
        <v-title title="Vue 组件化"></v-title>
        <v-button @click="handleClick">点击按钮</v-button>
    </div>
</template>
<script>
    // 导入组件
    import vTitle from './title.vue';
    import vButton from './button.vue';

    export default {
        components: {
            vTitle,
            vButton
        },
        methods: {
            handleClick (e) {
                console.log(e);
```

```
            }
        }
    }
</script>
```

ES 6 语法提示：

```
components: {
    vTitle,
    vButton
}
```

等同于：

```
components: {
    vTitle: vTitle,
    vButton: vButton
}
```

对象字面量缩写。当对象的 key 和 value 名称一致时，可以缩写成一个。

导入的组件都是局部注册的，而且可以自定义名称，其他用法和组件一致。

打开浏览器，如果已经正确渲染出了这两个组件，那么恭喜你，已经进入 Vue.js 的高级领域了，可以更高效地开发 Vue 项目，后面的章节都会基于 webpack 和单文件组件展开。

10.4 用于生产环境

我们先对 webpack 进一步配置，来支持更多常用的功能。

安装 url-loader 和 file-loader 来支持图片、字体等文件：

```
npm install --save-dev url-loader
npm install --save-dev file-loader
```

```
// webpack.config.js
var config = {
    // ...
    module: {
        rules: [
            // ...
            {
                test: /\.(gif|jpg|png|woff|svg|eot|ttf)\??.*$/,
                loader: 'url-loader?limit=1024'
            }
        ]
```

        }
    };

当遇到 .gif、.png、.ttf 等格式文件时，url-loader 会把它们一起编译到 dist 目录下，"?limit=1024" 是指如果这个文件小于 1kb，就以 base64 的形式加载，不会生成一个文件。

找一张图片，保存为 demo/images/image.png，并在 app.vue 中加载它：

```
<template>
    <div>
        <v-title title="Vue 组件化"></v-title>
        <v-button @click="handleClick">点击按钮</v-button>
        <p>
            <img src="./images/image.png" style="width: 200px;">
        </p>
    </div>
</template>
```

效果如图 10-6 所示。

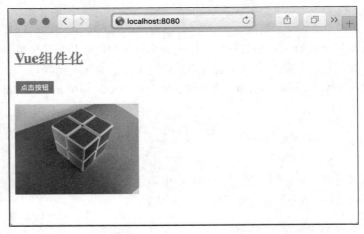

图 10-6　在 webpack 项目中使用图片

介绍打包上线前，先来分析 webpack 打包后的产物有哪些。

本书所介绍和使用的都是单页面富应用（SPA）技术，这意味着最终只有一个 html 的文件，其余都是静态资源。实际部署到生产环境时，一般会将 html 挂在后端程序下，由后端路由渲染这个页面，将所有的静态资源（css、js、image、iconfont 等）单独部署到 CDN，当然也可以和后端程序部署在一起，这样就实现了前后端完全分离。

我们在 webpack 的 ouput 选项里已经指定了 path 和 publicPath，打完包后，所有的资源都会保存在 demo/dist 目录下。

打包会用到下面两个依赖，使用 NPM 安装：

```
npm install --save-dev webpack-merge
npm install --save-dev html-webpack-plugin
```

为了方便开发和生产环境的切换，我们在 demo 目录下再新建一个用于生产环境的配置文件 webpack.prod.config.js。

编译打包，直接执行 webpack 命令就可以。在 package.json 中，再加入一个 build 的快捷脚本用来打包：

```
"scripts": {
  "dev": "webpack-dev-server --open --config webpack.config.js",
  "build": "webpack --progress --hide-modules --config webpack.prod.config.js"
}
```

先来看一下 webpack.prod.config.js 的代码：

```
var webpack = require('webpack');
var HtmlwebpackPlugin = require('html-webpack-plugin');
var ExtractTextPlugin = require('extract-text-webpack-plugin');
var merge = require('webpack-merge');
var webpackBaseConfig = require('./webpack.config.js');

// 清空基本配置的插件列表
webpackBaseConfig.plugins = [];

module.exports = merge(webpackBaseConfig, {
    output: {
        publicPath: '/dist/',
        //将入口文件重命名为带有20位hash值的唯一文件
        filename: '[name].[hash].js'
    },
    plugins: [
        new ExtractTextPlugin({
            // 提取css，并重命名为带有20位hash值的唯一文件
            filename: '[name].[hash].css',
            allChunks: true
        }),
        // 定义当前node环境为生产环境
        new webpack.DefinePlugin({
            'process.env': {
                NODE_ENV: '"production"'
            }
        }),
        // 压缩js
        new webpack.optimize.UglifyJsPlugin({
            compress: {
                warnings: false
```

```
            }
        }),
        // 提取模板,并保存入口 html 文件
        new HtmlWebpackPlugin({
            filename: '../index_prod.html',
            template: './index.ejs',
            inject: false
        })
    ]
});
```

上面安装的 webpack-merge 模块就是用于合并两个 webpack 的配置文件,所以 prod 的配置是在 webpack.config.js 基础上扩展的。静态资源在大部分场景下都有缓存(304),更新上线后一般都希望用户能及时地看到内容,所以给打包后的 css 和 js 文件的名称都加了 20 位的 hash 值,这样文件名就唯一了(比如 main.b3dd20e2dae9d76af86b.js),只要不对 html 文件设置缓存,上线后立即就可以加载最新的静态资源。

html-webpack-plugin 是用来生成 html 文件的,它通过 template 选项来读取指定的模板 index.ejs,然后输出到 filename 指定的目录,也就是 demo/index_prod.html。模板 index.ejs 动态设置了静态资源的路径和文件名,代码如下:

```
<!DOCTYPE html>
<html lang="zh-CN">
<head>
    <meta charset="UTF-8">
    <title>webpack App</title>
    <link rel="stylesheet" href="<%= htmlwebpackPlugin.files.css[0] %>">
</head>
<body>
    <div id="app"></div>
    <script type="text/javascript" src="<%= htmlwebpackPlugin.files.js[0] %>"></script>
</body>
</html>
```

ejs 是一个 JavaScript 模板库,用来从 JSON 数据中生成 HTML 字符串,常用于 Node.js。

最后在终端运行 npm run build,等待一会就会完成打包,成功后在 demo 下会生成一个 dist 的目录,里面就是打包完的所有静态资源。打包过程如图 10-7 所示。

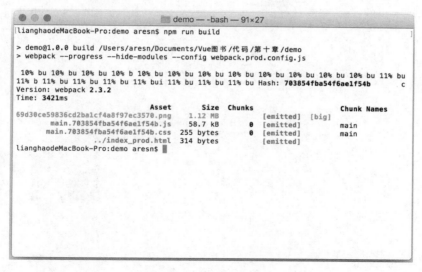

图 10-7　打包过程

以上就是 webpack 的核心功能和主要配置。除了本章介绍的这些内容外，webpack 还有很多高级的配置和丰富的插件及加载器，读者可查阅 webpack 文档进一步学习：https://webpack.js.org/。本章所有的代码已上传至 GitHub，访问下面的链接可以查看到并直接使用：

https://github.com/icarusion/vue-book

vue-book 下的 demo 目录就是本章的代码，在该目录下执行 npm install 命令会自动安装所有的依赖，然后执行 npm run dev 启动服务。

第 11 章

插　件

Vue.js 提供了插件机制,可以在全局添加一些功能。它们可以简单到几个方法、属性,也可以很复杂,比如一整套组件库。本章将介绍几个官方的核心插件,然后通过实战来开发一个插件。

注册插件需要一个公开的方法 install,它的第一个参数是 Vue 构造器,第二个参数是一个可选的选项对象。示例代码如下:

```
MyPlugin.install = function (Vue, options) {
    // 全局注册组件（指令等功能资源类似）
    Vue.component('component-name', {
        //组件内容
    })
    //添加实例方法
    Vue.prototype.$Notice = function () {
        // 逻辑...
    }
    //添加全局方法或属性
    Vue.globalMethod = function () {
        // 逻辑...
    }
    //添加全局混合
    Vue.mixin({
        mounted: function () {
            // 逻辑...
        }
    })
}
```

通过 Vue.use()来使用插件：

```
Vue.use(MyPlugin)
//或
Vue.use(MyPlugin, {
    // 参数
})
```

绝大多数情况下，开发插件主要是通过 NPM 发布后给别人使用的，在自己的项目中可以直接在入口调用以上的方法，无须多一步注册和使用的步骤。

 本章示例是基于上一章的 webpack 配置开始的，会在此基础上进一步开发，所以要确保先正确跑通了上一章的示例，也可以直接从 https://github.com/icarusion/vue-book 下载最终的示例。

11.1 前端路由与 vue-router

11.1.1 什么是前端路由

上一章介绍 webpack 时提到它的主要使用场景是单页面富应用（SPA），而 SPA 的核心就是前端路由。那什么是路由呢？通俗地讲，就是网址，比如 https://www.iviewui.com/docs/guide/introduce；再专业一点，就是每次 GET 或者 POST 等请求在服务端有一个专门的正则配置列表，然后匹配到具体的一条路径后，分发到不同的 Controller，进行各种操作，最终将 html 或数据返回给前端，这就完成了一次 IO。

当然，目前绝大多数的网站都是这种后端路由，也就是多页面的，这样的好处有很多，比如页面可以在服务端渲染好直接返回给浏览器，不用等待前端加载任何 js 和 css 就可以直接显示网页内容，再比如对 SEO 的友好等。后端路由的缺点也是很明显的，就是模板是由后端来维护或改写的。前端开发者需要安装整套的后端服务，必要时还得学习像 PHP 或 Java 这些非前端语言来改写 html 结构，所以 html 和数据、逻辑混为一谈，维护起来既臃肿又麻烦。

然后就有了前后端分离的开发模式，后端只提供 API 来返回数据，前端通过 Ajax 获取数据后，再用一定的方式渲染到页面里，这么做的优点就是前后端做的事情分得很清楚，后端专注在数据上，前端专注在交互和可视化上，如果今后再开发移动 App，那就正好能使用一套 API 了。当然，缺点也很明显，就是首屏渲染需要时间来加载 css 和 js。这种开发模式被很多公司认同，也出现了很多前端技术栈，比如 jQuery + artTemplate + Seajs(requirejs) + gulp 为主的开发模式可谓是万金油了。在 Node.js 出现后，这种现象有了改善，就是所谓的大前端，得益于 Node.js 和 JavaScript 的语言特性，html 模板可以完全由前端来控制，同步或异步渲染完全由前端自由决定，并且由前端维护一套模板，这就是为什么在服务端使用 artTemplate、React 以及 Vue 2 的原因。说了这么多，到底怎样算是 SPA 呢？其实就是在前后端分离的基础上，加一层前端路由。

前端路由，即由前端来维护一个路由规则。实现有两种，一种是利用 url 的 hash，就是常说的锚点（#），JavaScript 通过 hashChange 事件来监听 url 的改变，IE7 及以下需要用轮询；另一种就是 HTML5 的 History 模式，它使 url 看起来像普通网站那样，以 "/" 分割，没有#，但页面并没有跳转，不过使用这种模式需要服务端支持，服务端在接收到所有的请求后，都指向同一个 html 文件，不然会出现 404。因此，SPA 只有一个 html，整个网站所有的内容都在这一个 html 里，通过 JavaScript 来处理。

前端路由的优点有很多，比如页面持久性，像大部分音乐网站，你都可以在播放歌曲的同时跳转到别的页面，而音乐没有中断。再比如前后端彻底分离。前端路由的框架通用的有 Director（https://github.com/flatiron/director），不过更多还是结合具体框架来用，比如 Angular 的 ngRouter，React 的 ReactRouter，以及本节要介绍的 Vue 的 vue-router。

如果要独立开发一个前端路由，需要考虑到页面的可插拔、页面的生命周期、内存管理等问题。

11.1.2 vue-router 基本用法

回顾第 7 章 7.5.3 小节，当时介绍了通过 is 特性来实现动态组件的方法。vue-router 的实现原理与之类似，路由不同的页面事实上就是动态加载不同的组件。

新建一个目录 router，复制上一章的代码并安装完成后，再通过 NPM 来安装 vue-router：

```
npm install --save vue-router
```

在 main.js 里使用 Vue.use() 加载插件：

```
import Vue from 'vue';
import VueRouter from 'vue-router';
import App from './app.vue';

Vue.use(VueRouter);
```

每个页面对应一个组件，也就是对应一个 .vue 文件。在 router 目录下创建 views 目录，用于存放所有的页面，然后在 views 里创建 index.vue 和 about.vue 两个文件：

```
// index.vue
<template>
    <div>首页</div>
</template>
<script>
    export default {

    }
</script>


// about.vue
<template>
    <div>介绍页</div>
</template>
<script>
```

```
    export default {

    }
</script>
```

再回到 main.js 里,完成路由的剩余配置。创建一个数组来制定路由匹配列表,每一个路由映射一个组件:

```
const Routers = [
    {
        path: '/index',
        component: (resolve) => require(['./views/index.vue'], resolve)
    },
    {
        path: '/about',
        component: (resolve) => require(['./views/about.vue'], resolve)
    }
];
```

ES 6 语法提示:

在 ES 6 中,使用 let 和 const 命令来声明变量,代替了 var。let 和 const 的作用域是"块",比如:

```
    {
        let a = 1;
        var b = 2;
    }

    console.log(b);    // 2
    console.log(a);    // 报错: a is not defined
```

const 与 let 的主要区别是,const 用于声明常量,也就是声明后不能再修改。

如果一时还不了解它们的其他区别,可以先把 let 和 const 当作 var 来理解。

Routers 里每一项的 path 属性就是指定当前匹配的路径,component 是映射的组件。上例的写法,webpack 会把每一个路由都打包为一个 js 文件,在请求到该页面时,才去加载这个页面的 js,也就是异步实现的懒加载(按需加载),这与第 7 章 7.5.4 小节异步组件的用法类似。这样做的好处是不需要在打开首页的时候就把所有的页面内容全部加载进来,只在访问时才加载。如果非要一次性加载,可以这样写:

```
{
    path: '/index',
    component: require('./views/index.vue')
}
```

使用了异步路由后，编译出的每个页面的 js 都叫作 chunk（块），它们命名默认是 0.main.js、1.main.js ……可以在 webpack 配置的出口 output 里通过设置 chunkFilename 字段修改 chunk 命名，例如：

```
output: {
    publicPath: '/dist/',
    filename: '[name].js',
    chunkFilename: '[name].chunk.js'
}
```

有了 chunk 后，在每个页面(.vue 文件)里写的样式也需要配置后才会打包进 main.css，否则仍然会通过 JavaScript 动态创建<style> 标签的形式写入。配置插件：

```
// webpack.config.js
plugins: [
    new ExtractTextPlugin({
        filename: '[name].css',
        allChunks: true
    })
]
```

然后继续在 main.js 里完成配置和路由实例：

```
const RouterConfig = {
    // 使用 HTML5 的 History 路由模式
    mode: 'history',
    routes: Routers
};
const router = new VueRouter(RouterConfig);

new Vue({
    el: '#app',
    router: router,
    render: h => {
        return h(App)
    }
});
```

在 RouterConfig 里，设置 mode 为 history 会开启 HTML 5 的 History 路由模式，通过 "/" 设置路径。如果不配置 mode，就会使用 "#" 来设置路径。开启 History 路由，在生产环境时服务端必须进行配置，将所有路由都指向同一个 html，或设置 404 页面为该 html，否则刷新时页面会出现 404。

webpack-dev-server 也要配置下来支持 History 路由，在 package.json 中修改 dev 命令：

```
"scripts": {
    "dev": "webpack-dev-server --open --history-api-fallback --config webpack.config.js"
}
```

增加了--history-api-fallback，所有的路由都会指向 index.html。

配置好了这些，最后在根实例 app.vue 里添加一个路由视图<router-view>来挂载所有的路由组件：

```
<template>
    <div>
        <router-view></router-view>
    </div>
</template>
<script>
    export default {

    }
</script>
```

运行网页时，<router-view>会根据当前路由动态渲染不同的页面组件。网页中一些公共部分，比如顶部的导航栏、侧边导航栏、底部的版权信息，这些也可以直接写在 app.vue 里，与<router-view>同级。路由切换时，切换的是<router-view>挂载的组件，其他的内容并不会变化。

运行 npm run dev 启动服务，然后访问 127.0.0.1:8080/index 和 127.0.0.1:8080/about 就可以访问这两个页面了。

在路由列表里，可以在最后新加一项，当访问的路径不存在时，重定向到首页：

```
const Routers = [
    // ...
    {
        path: '*',
        redirect: '/index'
    }
];
```

这样直接访问 127.0.0.1:8080，就自动跳转到了 127.0.0.1:8080/index。

路由列表的 path 也可以带参数，比如"个人主页"的场景，路由的一部分是固定的，一部分是动态的：/user/123456，其中用户 id "123456" 就是动态的，但它们路由到同一个页面，在这个页面里，期望获取这个 id，然后请求相关数据。在路由里可以这样配置参数：

```
// main.js
const Routers = [
    // ...
    {
        path: '/user/:id',
        component: (resolve) => require(['./views/user.vue'], resolve)
```

```
    },
    {
        path: '*',
        redirect: '/index'
    }
];
//在 router/views 目录下，新建 user.vue 文件
<template>
    <div>{{ $route.params.id }}</div>
</template>
<script>
    export default {
        mounted () {
            console.log(this.$route.params.id);
        }
    }
</script>
```

这里的 this.$route 可以访问到当前路由的很多信息，可以打印出来看看都有什么，在开发中会经常用到里面的数据。

因为配置的路由是"/user/:id"，所以直接访问 127.0.0.1:8080/user 会重定向到/index，需要带一个 id 才能到 user.vue，比如 127.0.0.1:8080/user/123456。

11.1.3 跳转

vue-router 有两种跳转页面的方法，第一种是使用内置的<router-link>组件，它会被渲染为一个<a>标签：

```
// index.vue
<template>
    <div>
        <h1>首页</h1>
        <router-link to="/about">跳转到 about</router-link>
    </div>
</template>
```

它的用法与一般的组件一样，to 是一个 prop，指定需要跳转的路径，当然也可以用 v-bind 动态设置。使用<router-link>，在 HTML5 的 History 模式下会拦截点击，避免浏览器重新加载页面。

<router-link> 还有其他的一些 prop，常用的有：

- tag
  tag 可以指定渲染成什么标签，比如<router-link to="/about" tag="li"> 渲染的结果就是<li>而不是<a>。

- replace

  使用 replace 不会留下 History 记录，所以导航后不能用后退键返回上一个页面，如 `<router-link to="/about" replace>`。

- active-class

  当`<router-link>`对应的路由匹配成功时，会自动给当前元素设置一个名为 router-link-active 的 class，设置 prop: active-class 可以修改默认的名称。在做类似导航栏时，可以使用该功能高亮显示当前页面对应的导航菜单项，但是一般不会修改 active-class，直接使用默认值 router-link-active 就可以。

有时候，跳转页面可能需要在 JavaScript 里进行，类似于 window.location.href。这时可以用第二种跳转方法，使用 router 实例的方法。比如在 about.vue 里，通过点击事件跳转：

```
// about.vue
<template>
    <div>
        <h1>介绍页</h1>
        <button @click="handleRouter">跳转到 user</button>
    </div>
</template>
<script>
    export default {
        methods: {
            handleRouter () {
                this.$router.push('/user/123');
            }
        }
    }
</script>
```

$router 还有其他一些方法：

- replace

  类似于`<router-link>`的 replace 功能，它不会向 history 添加新记录，而是替换掉当前的 history 记录，如 this.$router.replace('/user/123');。

- go

  类似于 window.history.go()，在 history 记录中向前或者后退多少步，参数是整数，例如：

```
// 后退 1 页
this.$router.go(-1);
// 前进 2 页
this.$router.go(2);
```

11.1.4 高级用法

本节将从实际业务需求出发，逐步探索 vue-router 的高级用法。

先抛出一个问题：在 SPA 项目中，如何修改网页的标题？

网页标题是通过<title></title>来显示的，但是 SPA 只有一个固定的 html，切换到不同页面时，标题并不会变化，但是可以通过 JavaScript 修改 <title>的内容：

```
window.document.title = '要修改的网页标题';
```

那么问题就来了，在 Vue 工程里，在哪里、在什么时候修改标题呢？比较容易想到的一个办法是，在每个页面的.vue 文件里，通过 mounted 钩子修改。这种办法没有问题，但是页面多了维护起来会很麻烦，而且这些逻辑都是重复的。

比较理想的一个思路就是，在页面发生路由改变时，统一设置。vue-router 提供了导航钩子 beforeEach 和 afterEach，它们会在路由即将改变前和改变后触发，所以设置标题可以在 beforeEach 钩子完成。

```
// main.js
const Routers = [
    {
        path: '/index',
        meta: {
            title: '首页'
        },
        component: (resolve) => require(['./views/index.vue'], resolve)
    },
    {
        path: '/about',
        meta: {
            title: '关于'
        },
        component: (resolve) => require(['./views/about.vue'], resolve)
    },
    {
        path: '/user/:id',
        meta: {
            title: '个人主页'
        },
        component: (resolve) => require(['./views/user.vue'], resolve)
    },
    {
        path: '*',
        redirect: '/index'
    }
];

const router = new VueRouter(RouterConfig);
router.beforeEach((to, from, next) => {
```

```
    window.document.title = to.meta.title;
    next();
});
```

导航钩子有 3 个参数：

- to   即将要进入的目标的路由对象。
- from   当前导航即将要离开的路由对象。
- next   调用该方法后，才能进入下一个钩子。

路由列表的 meta 字段可以自定义一些信息，比如我们将每个页面的 title 写入了 meta 来统一维护，beforeEach 钩子可以从路由对象 to 里获取 meta 信息，从而改变标题。

有了这两个钩子，还能做很多事情来提升用户体验。比如一个页面较长，滚动到某个位置，再跳转到另一个页面，滚动条默认是在上一个页面停留的位置，而好的体验肯定是能返回顶端。通过钩子 afterEach 就可以实现：

```
// main.js
// ...
router.afterEach((to, from, next) => {
    window.scrollTo(0, 0);
});
```

类似的需求还有，从一个页面过渡到另一个页面时，可以出现一个全局的 Loading 动画，等到新页面加载完后再结束动画。

next()方法还可以设置参数，比如下面的场景。

某些页面需要校验是否登录，如果登录了就可以访问，否则跳转到登录页。这里我们通过 localStorage 来简易判断是否登录，示例代码如下：

```
router.beforeEach((to, from, next) => {
    if (window.localStorage.getItem('token')) {
        next();
    } else {
        next('/login');
    }
});
```

next()的参数设置为 false 时，可以取消导航，设置为具体的路径可以导航到指定的页面。

正确地使用好导航钩子可以方便实现一些全局的功能，而且便于维护。更多的可能需要在业务中不断探索。

本节所有的代码已上传至 GitHub，访问下面的链接可以查看到并直接使用：

https://github.com/icarusion/vue-book

vue-book 下的 router 目录就是本节的代码，在该目录下执行 npm install 命令会自动安装所有的依赖，然后执行 npm run dev 启动服务。

拓展阅读建议：vue-router 还有一些不常用或可不用的功能，比如嵌套路由、路由的命名、视图的命名等，读者可以查阅文档进一步学习，网址为 https://router.vuejs.org/。

11.2 状态管理与 Vuex

11.2.1 状态管理与使用场景

回顾第 7 章的 7.3.3 小节，我们在介绍非父子组件（也就是跨级组件和兄弟组件）通信时，使用了 bus（中央事件总线）的一个方法，用来触发和接收事件，进一步起到通信的作用。Vuex 所解决的问题与 bus 类似，它作为 Vue 的一个插件来使用，可以更好地管理和维护整个项目的组件状态。

一个组件可以分为数据（model）和视图（view），数据更新时，视图也会自动更新。在视图中又可以绑定一些事件，它们触发 methods 里指定的方法，从而可以改变数据、更新视图，这是一个组件基本的运行模式。比如下面的示例：

```
// message.vue
<template>
    <div>
        {{ message }}
        <button @click="handleClick">Change word</button>
    </div>
</template>
<script>
    export default {
        data () {
            return {
                message: 'Hello World.'
            };
        },
        methods: {
            handleClick () {
                this.message = 'Hello Vue.';
            }
        }
    };
</script>
```

这里的数据 message 和方法 handleClick 只有在 message.vue 组件里可以访问和使用，其他的组件是无法读取和修改 message 的。但是在实际业务中，经常有跨组件共享数据的需求，因此 Vuex 的设计就是用来统一管理组件状态的，它定义了一系列规范来使用和操作数据，使组件应用更加高效。

使用 Vuex 会有一定的门槛和复杂性，它的主要使用场景是大型单页应用，更适合多人协同开发。如果你的项目不是很复杂,或者希望短期内见效,你需要认真考虑是否真的有必要使用 Vuex,也许像 7.3.3 小节介绍的 bus 方法就能很简单地解决你的需求。当然,并不是所有大型多人协同开发的 SPA 项目都必须使用 Vuex,事实上,我们在一些生产环境中只是使用 bus 也能实现得很好,用与否主要取决于你的团队和技术储备。

每一个框架的诞生都是用来解决具体问题的。虽然 bus 已经可以很好地解决跨组件通信,但它在数据管理、维护、架构设计上还只是一个简单的组件,而 Vuex 却能更优雅和高效地完成状态管理。

11.2.2　Vuex 基本用法

本节是在上一节的 vue-router 基础之上进行开发的,在本地创建目录 vuex,然后复制上一节的所有代码,或直接从 https://github.com/icarusion/vue-book 下载后,使用 router 目录下的代码。

首先通过 NPM 安装 Vuex:

```
npm install --save vuex
```

它的用法与 vue-router 类似,在 main.js 里,通过 Vue.use() 使用 Vuex:

```
import Vue from 'vue';
import VueRouter from 'vue-router';
import Vuex from 'vuex';
import App from './app.vue';

Vue.use(VueRouter);
Vue.use(Vuex);

// 路由配置
// 省略...

const store = new Vuex.Store({
    // vuex 的配置
});

new Vue({
    el: '#app',
    router: router,
    // 使用 vuex
    store: store,
    render: h => {
        return h(App)
    }
});
```

仓库 store 包含了应用的数据(状态)和操作过程。Vuex 里的数据都是响应式的,任何组件使用同一 store 的数据时,只要 store 的数据变化,对应的组件也会立即更新。

数据保存在 Vuex 选项的 state 字段内,比如要实现一个计数器,定义一个数据 count,初始值为 0:

```js
const store = new Vuex.Store({
    state: {
        count: 0
    }
});
```

在任何组件内,可以直接通过 $store.state.count 读取:

```html
// index.vue
<template>
    <div>
        <h1>首页</h1>
        {{ $store.state.count }}
    </div>
</template>
```

直接写在 template 里显得有点乱,可以用一个计算属性来显示:

```html
<template>
    <div>
        <h1>首页</h1>
        {{ count }}
    </div>
</template>
<script>
    export default {
        computed: {
            count () {
                return this.$store.state.count;
            }
        }
    }
</script>
```

现在访问首页,计数 0 已经可以显示出来了。

在组件内,来自 store 的数据只能读取,不能手动改变,改变 store 中数据的唯一途径就是显式地提交 mutations。

mutations 是 Vuex 的第二个选项,用来直接修改 state 里的数据。我们给计数器增加 2 个 mutations,用来加 1 和减 1:

```js
// main.js
const store = new Vuex.Store({
    state: {
```

```
        count: 0
    },
    mutations: {
        increment (state) {
            state.count ++;
        },
        decrease (state) {
            state.count --;
        }
    }
});
```

在组件内,通过 this.$store.commit 方法来执行 mutations。在 index.vue 中添加两个按钮用于加和减:

```
<template>
    <div>
        <h1>首页</h1>
        {{ count }}
        <button @click="handleIncrement">+1</button>
        <button @click="handleDecrease">-1</button>
    </div>
</template>
<script>
    export default {
        computed: {
            count () {
                return this.$store.state.count;
            }
        },
        methods: {
            handleIncrement () {
                this.$store.commit('increment');
            },
            handleDecrease () {
                this.$store.commit('decrease');
            }
        }
    }
</script>
```

这看起来很像 JavaScript 的观察者模式,组件只负责提交一个事件名,Vuex 对应的 mutations 来完成业务逻辑。

mutations 还可以接受第二个参数,可以是数字、字符串或对象等类型。比如每次增加的不是 1,而是指定的数量,可以这样改写:

```js
// main.js,部分代码省略
mutations: {
    increment (state, n = 1) {
        state.count += n;
    }
}
```

ES 6 语法提示:
函数的参数可以设定默认值,当没有传入该参数时,使用设置的值。比如上例的 increment (state, n = 1)等同于:

```js
increment (state, n) {
    n = n || 1;
}
```

```html
// index.vue,部分代码省略
<template>
    <div>
        <button @click="handleIncrementMore">+5</button>
    </div>
</template>
<script>
    export default {
        methods: {
            handleIncrementMore () {
                this.$store.commit('increment', 5);
            }
        }
    }
</script>
```

当一个参数不够用时,可以传入一个对象,无限扩展。

提交 mutation 的另一种方式是直接使用包含 type 属性的对象,比如:

```js
// main.js
mutations: {
    increment (state, params) {
        state.count += params.count;
    }
}
```

```
// index.vue
this.$store.commit({
    type: 'increment',
    count: 10
});
```

注意，mutation 里尽量不要异步操作数据。如果异步操作数据了，组件在 commit 后，数据不能立即改变，而且不知道什么时候会改变。在下一节的 actions 里会介绍如何处理异步。

11.2.3 高级用法

Vuex 还有其他 3 个选项可以使用：getters、actions、modules。

有这样的一个场景：Vuex 定义了某个数据 list，它是一个数组，比如：

```
// main.js，部分代码省略
const store = new Vuex.Store({
    state: {
        list: [1, 5, 8, 10, 30, 50]
    }
});
```

如果只想得到小于 10 的数据，最容易想到的方法可能是在组件的计算属性里进行过滤。示例代码如下：

```
// index.vue，部分代码省略
<template>
    <div>
        <div>{{ list }}</div>
    </div>
</template>
<script>
    export default {
        computed: {
            list () {
                return this.$store.state.list.filter(item => item < 10);
            }
        }
    }
</script>
```

这样写完全没有问题。但如果还有其他的组件也需要过滤后的数据时，就得把 computed 的代码完全复制一份，而且需要修改过滤方法时，每个用到的组件都得修改，这明显不是我们期望的结果。如果能将 computed 的方法也提取出来就方便多了，getters 就是来做这件事的。

使用 getters 改写上面的示例：

```js
// main.js,部分代码省略
const store = new Vuex.Store({
    state: {
        list: [1, 5, 8, 10, 30, 50]
    },
    getters: {
        filteredList: state => {
            return state.list.filter(item => item < 10);
        }
    }
});
```

```html
// index.vue,部分代码省略
<template>
    <div>
        <div>{{ list }}</div>
    </div>
</template>
<script>
    export default {
        computed: {
            list () {
                return this.$store.getters.filteredList;
            }
        }
    }
</script>
```

这种用法与组件的计算属性非常像。getter 也可以依赖其他的 getter,把 getter 作为第二个参数。比如再写一个 getter,计算出 list 过滤后的结果的数量:

```js
// main.js
const store = new Vuex.Store({
    state: {
        list: [1, 5, 8, 10, 30, 50]
    },
    getters: {
        filteredList: state => {
            return state.list.filter(item => item < 10);
        },
        listCount: (state, getters) => {
            return getters.filteredList.length;
        }
    }
```

```
});

// index.vue
<template>
    <div>
        <div>{{ list }}</div>
        <div>{{ listCount }}</div>
    </div>
</template>
<script>
    export default {
        computed: {
            list () {
                return this.$store.getters.filteredList;
            },
            listCount () {
                return this.$store.getters.listCount;
            }
        }
    }
</script>
```

上一节提到，mutation 里不应该异步操作数据，所以有了 actions 选项。action 与 mutation 很像，不同的是 action 里面提交的是 mutation，并且可以异步操作业务逻辑。

action 在组件内通过 $store.dispatch 触发，例如使用 action 来加 1：

```
// main.js 部分代码省略
const store = new Vuex.Store({
    state: {
        count: 0
    },
    mutations: {
        increment (state, n = 1) {
            state.count += n;
        }
    },
    actions: {
        increment (context) {
            context.commit('increment');
        }
    }
});

// index.vue 部分代码省略
```

```html
<template>
    <div>
        {{ count }}
        <button @click="handleActionIncrement">action +1</button>
    </div>
</template>
<script>
    export default {
        computed: {
            count () {
                return this.$store.state.count;
            }
        },
        methods: {
            handleActionIncrement () {
                this.$store.dispatch('increment');
            }
        }
    }
</script>
```

是不是觉得有点多此一举？没错，就目前示例来看的确是，因为可以直接在组件 commit mutation，没必要通过 action 中转一次。但是加了异步就不一样了，我们用一个 Promise 在 1 秒钟后提交 mutation，示例代码如下：

```
// main.js 部分代码省略
const store = new Vuex.Store({
    state: {
        count: 0
    },
    mutations: {
        increment (state, n = 1) {
            state.count += n;
        }
    },
    actions: {
        asyncIncrement (context) {
            return new Promise(resolve => {
                setTimeout(() => {
                    context.commit('increment');
                    resolve();
                }, 1000)
            });
        }
```

```
    }
});

// index.vue 部分代码省略
<template>
    <div>
        {{ count }}
        <button @click="handleAsyncIncrement">async +1</button>
    </div>
</template>
<script>
    export default {
        computed: {
            count () {
                return this.$store.state.count;
            }
        },
        methods: {
            handleAsyncIncrement () {
                this.$store.dispatch('asyncIncrement').then(() => {
                    console.log(this.$store.state.count); // 1
                });
            }
        }
    }
</script>
```

ES 6 语法提示：

Promise 是一种异步方案，它有 3 种状态：Pending（进行中）、Resolved（已完成）、Rejected（已失败）。比如下面的示例，通过判断一个随机数是否大于 0.5 来模拟完成与失败：

```
const promise = new Promise((resolve, reject) => {
    setTimeout(() => {
        const random = Math.random();
        if (random > 0.5) {
            resolve(random);
        } else {
            reject(random);
        }
    }, 1000);
});
```

```js
promise.then((value) => {
    console.log('success', value);
}).catch((error) => {
    console.log('fail', error);
});
```

如果暂时还不理解 Promise，异步 action 的示例还可以用普通的回调来实现：

```js
// main.js
actions: {
    asyncIncrement (context, callback) {
        setTimeout(() => {
            context.commit('increment');
            callback();
        }, 1000);
    }
}

// index.vue
methods: {
    handleAsyncIncrement () {
        this.$store.dispatch('asyncIncrement', () => {
            console.log(this.$store.state.count); // 1
        });
    }
}
```

mutations、actions 看起来很相似，可能会觉得不知道该用哪个，但是 Vuex 很像是一种与开发者的约定，所以涉及改变数据的，就使用 mutations，存在业务逻辑的，就用 actions。至于将业务逻辑放在 action 里还是 Vue 组件里完成，就需要根据实际场景拿捏了。

最后一个选项是 modules，它用来将 store 分割到不同模块。当你的项目足够大时，store 里的 state、getters、mutations、actions 会非常多，都放在 main.js 里显得不是很友好，使用 modules 可以把它们写到不同的文件中。每个 module 拥有自己的 state、getters、mutations、actions，而且可以多层嵌套。

比如下面的示例：

```js
const moduleA = {
    state: { ... },
    mutations: { ... },
    actions: { ... },
    getters: { ... }
}

const moduleB = {
```

```
    state: { ... },
    mutations: { ... },
    actions: { ... }
}

const store = new Vuex.Store({
    modules: {
        a: moduleA,
        b: moduleB
    }
})

store.state.a // moduleA 的状态
store.state.b // moduleB 的状态
```

module 的 mutation 和 getter 接收的第一个参数 state 是当前模块的状态。在 actions 和 getters 中，还可以接收一个参数 rootState，来访问根节点的状态。比如 getters 中 rootState 将作为第 3 个参数：

```
const moduleA = {
    state: {
        count: 0
    },
    getters: {
        sumCount (state, getters, rootState) {
            return state.count + rootState.count;
        }
    }
}
```

本节所有的代码已上传至 GitHub，访问下面的链接可以查看到并直接使用：

https://github.com/icarusion/vue-book

vue-book 下的 vuex 目录就是本节的代码，在该目录下执行 npm install 命令会自动安装所有的依赖，然后执行 npm run dev 启动服务。

拓展阅读建议：可以到 Vuex 文档进一步阅读它更多的用法：https://vuex.vuejs.org 。

11.3 实战：中央事件总线插件 vue-bus

在第 7 章的 7.3.3 小节介绍了中央事件总线 bus 的用法，它作为一个简单的组件传递事件，用于解决跨级和兄弟组件通信的问题。本节将使用该思想将其封装为一个 Vue 的插件，可以在所有组件间随意使用，而不需要导入 bus。

本节是在上一节的 Vuex 基础之上进行开发的,在本地创建目录 vue-bus,然后复制上一节的所有代码,或直接从 https://github.com/icarusion/vue-book 下载后,使用 vuex 目录下的代码。

完成基本安装后,在 vue-bus 目录下新建 vue-bus.js 文件。vue-bus 插件像 vue-router 和 Vuex 一样,给 Vue 添加一个属性$bus,并代理 $emit、$on、$off 三个方法。代码如下:

```
// vue-bus.js
const install = function (Vue) {
    const Bus = new Vue({
        methods: {
            emit (event, ...args) {
                this.$emit(event, ...args);
            },
            on (event, callback) {
                this.$on(event, callback);
            },
            off (event, callback) {
                this.$off(event, callback);
            }
        }
    });
    Vue.prototype.$bus = Bus;
};
export default install;
```

ES 6 语法提示:
emit (event, ...args)中的...args 是函数参数的解构。因为不知道组件会传递多少个参数进来,使用...args 可以把从当前参数(这里是第二个)到最后的参数都获取到。

在 main.js 中使用插件:

```
// main.js,部分代码省略

import VueBus from './vue-bus';
Vue.use(VueBus);
```

在 views 目录下新建一个组件 Counter.vue:

```
// views/counter.vue
<template>
    <div>
        {{ number }}
        <button @click="handleAddRandom">随机增加</button>
    </div>
</template>
<script>
```

```
    export default {
        props: {
            number: {
                type: Number
            }
        },
        methods: {
            handleAddRandom () {
                // 随机获取 1~100 中的数
                const num = Math.floor(Math.random () * 100 + 1);
                this.$bus.emit('add', num);
            }
        }
    };
</script>
```

在 index.vue 中使用 Counter 组件并监听来自 counter.vue 的自定义事件：

```
// index.vue, 部分代码省略
<template>
    <div>
        随机增加：
        <Counter :number="number"></Counter>
    </div>
</template>
<script>
    import Counter from './counter.vue';

    export default {
        components: {
            Counter
        },
        data () {
            return {
                number: 0
            }
        },
        created () {
            this.$bus.on('add', num => {
                this.number += num;
            });
        }
    }
</script>
```

vue-bus 的代码比较简单，只有不到 20 行，但它却解决了跨组件通信的问题，而且通过插件的形式使用后，所有组件都可以直接使用$bus，而无须每个组件都导入 bus 组件。

使用 vue-bus 有两点需要注意，第一是$bus.on 应该在 created 钩子内使用，如果在 mounted 使用，它可能接收不到其他组件来自 created 钩子内发出的事件；第二点是使用了$bus.on，在 beforeDestroy 钩子里应该再使用$bus.off 解除，因为组件销毁后，就没必要把监听的句柄储存在 vue-bus 里了，所以 index.vue 可以适当改写为：

```vue
<template>
    <div>
        随机增加：
        <Counter :number="number"></Counter>
    </div>
</template>
<script>
    import Counter from './counter.vue';

    export default {
        components: {
            Counter
        },
        methods: {
            handleAddRandom (num) {
                this.number += num;
            }
        },
        data () {
            return {
                number: 0
            }
        },
        created () {
            this.$bus.on('add', this.handleAddRandom);
        },
        beforeDestroy () {
            this.$bus.off('add', this.handleAddRandom);
        }
    }
</script>
```

本节所有的代码已上传至 GitHub，访问下面的链接可以查看到并直接使用：

https://github.com/icarusion/vue-book

vue-book 下的 vue-bus 目录就是本节的代码，在该目录下执行 npm install 命令会自动安装所有的依赖，然后执行 npm run dev 启动服务。

练习：学习 XMLHttpRequest（即 XHR）相关知识，开发一个简单的$ajax 插件，用于异步获取服务端数据。以下实现 Ajax 的代码可以作为参考：

```
const ajax = function (options = {}) {
    options.type = (options.type || 'GET').toUpperCase();

    let data = [];
    for(let i in options.data){
        data.push(encodeURIComponent(i) +
            '=' + encodeURIComponent(options.data[i]));
    }
    data = data.join('&');

    const xhr = new XMLHttpRequest();
    xhr.onreadystatechange = function () {
        if (xhr.readyState === 4) {
            const status = xhr.status;
            if (status >= 200 && status < 300) {
                options.success &&
                options.success(JSON.parse(xhr.responseText),
xhr.responseXML);
            } else {
                options.error && options.error(status);
            }
        }
    };
    if (options.type === 'GET') {
        xhr.open('GET', options.url + '?' + data, true);
        xhr.send(null);
    } else if (options.type === 'POST') {
        xhr.open('POST', options.url, true);
        xhr.setRequestHeader(
            'Content-Type',
            'application/x-www-form-urlencoded');
        xhr.send(data);
    }
};
```

第 3 篇 实战篇

基础篇和进阶篇的内容已经涵盖了 Vue 2 绝大部分知识点,如果你已经认真阅读过,那么是时候实战了。从下一章开始,会从不同的几个维度来讲述一些实战案例。

第 12 章将介绍基于 Vue 2 的一套高质量 UI 组件库——iView。iView 是一整套的前端解决方案,本书会介绍它的几个具有代表性的组件设计思想并做代码剖析。

第 13 章将实现一个 Vue 版的知乎日报,会用到 webpack、vue-router、vuex。

第 14 章将开发一个简易的电商网站项目,包括商品列表、详情、购物车等常用功能,同样基于 webpack、vue-router 和 vuex。

实战篇的章节代码示例要比基础篇和进阶篇稍有难度,书写风格完全基于 ES 6,所以需要掌握进阶篇中所有的 ES 6 语法提示的内容。每一章的完整代码都提交在 GitHub,访问地址也会在章末提及。

实战篇的所有内容都是基于真实的生产环境考虑的,所涉及的内容涵盖 Vue.js 绝大部分 API,它很有借鉴意义,但并不一定是适合你的最佳实践。所以在你的项目中,可以根据实际情况来组织代码架构。比如你完全可以不用 ES 6,而只是使用 ES 5 就足以;再比如可以不使用 vuex,而用 bus 代替;当然,如果你的团队技术不错,可以使用 TypeScript 等语言来开发。总之,实战篇更多提供的是灵感,而你需要在技术选型上找到平衡。

第 12 章

iView 经典组件剖析

iView 是一套基于 Vue.js 2 的开源 UI 组件库,主要服务于 PC 界面的中后台产品。简单地理解,它是深度封装的 40 多个常用业务组件,比如 Input、Checkbox、Select、Table;但它同时也是一整套的前端解决方案,包括设计规范、基础样式,支持服务端渲染(SSR),同时提供了可视化脚手架,方便快速构建项目工程。iView 的官方网站和 GitHub 地址如下。

官方网站:https://www.iviewui.com/。

GitHub:https://github.com/iview/iview。

图 12-1 展示了部分 iView 组件的截图。

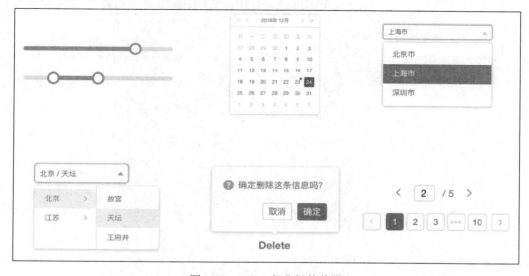

图 12-1 iView 部分组件截图

从左往右，从上到下依次为：Slider 滑块、DatePicker 日期选择器、Select 选择器、Cascader 级联选择器、Poptip 气泡提示、Page 分页。

iView 以高质量、细致漂亮的 UI、事无巨细的文档等特点成为 Vue.js 组件库中最受欢迎的项目之一。本章就来剖析 iView 几个具有代表性的组件，重点是理解设计一个通用组件和组件库的的思想和过程。

12.1　级联选择组件 Cascader

级联选择是网页应用中常见的表单类控件，主要用于省市区、公司级别、事务分类等关联数据集合的选择。iView 的级联选择组件如图 12-2 所示。

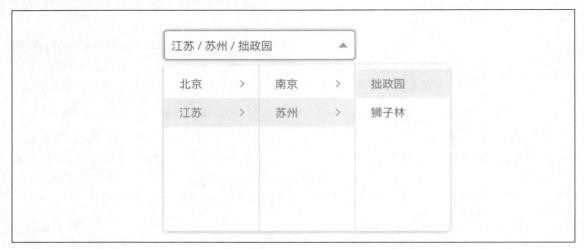

图 12-2　iView 级联选择组件

Cascader 接收一个 prop：data 作为选择面板的数据源，使用 v-model 可以双向绑定当前选择的项。比如图 12-2 的示例代码如下：

```
<template>
    <Cascader :data="data" v-model="value"></Cascader>
</template>
<script>
    export default {
        data () {
            return {
                value: ['jiangsu', 'suzhou', 'zhuozhengyuan'],
                data: [{
                    value: 'beijing',
                    label: '北京',
                    children: [
                        {
```

```
                    value: 'gugong',
                    label: '故宫'
                },
                {
                    value: 'tiantan',
                    label: '天坛'
                },
                {
                    value: 'wangfujing',
                    label: '王府井'
                }
            ]
        }, {
            value: 'jiangsu',
            label: '江苏',
            children: [
                {
                    value: 'nanjing',
                    label: '南京',
                    children: [
                        {
                            value: 'fuzimiao',
                            label: '夫子庙'
                        }
                    ]
                },
                {
                    value: 'suzhou',
                    label: '苏州',
                    children: [
                        {
                            value: 'zhuozhengyuan',
                            label: '拙政园'
                        },
                        {
                            value: 'shizilin',
                            label: '狮子林'
                        }
                    ]
                }
            ]
        }]
    }
```

```
        }
    }
</script>
```

data 中的 label 是面板显示的内容，value 是它对应的值，children 是它的子集，可递归。v-model 绑定一个数组，每一项对应 data 里的 value。

Cascader 对应的文档地址为 https://www.iviewui.com/components/cascader，源代码地址为 https://github.com/iview/iview/tree/2.0/src/components/cascader。

提示

iView 的每个组件往往依赖其他的组件，比如 Cascader 依赖 Input、Drop、Icon 组件，所以在剖析 Cascader 时，不会再对它依赖的组件进行详细介绍，可以到 GitHub 或文档查看依赖组件的相关内容。

本章是基于 iView 的 2.0.0-rc.10 版本，未来有可能会更新此组件，可到文档查阅更新日志：https://www.iviewui.com/docs/guide/update 。

本章的示例代码并不是完整的，只是剖析核心的功能，分析设计思路，完整的代码请前往 GitHub 查看。

开发一个通用组件最重要的是定义 API，Vue 组件的 API 来自 3 部分：prop、slot 和 event。API 决定了一个组件的所有功能，而且作为对外提供的组件，一旦 API 确定好后，如果再迭代更新，用户的代价就会很高，因为他们已经在业务中使用你的组件，改动太多意味着所有用到的地方都需要改动，所以组件库更新分兼容更新和不兼容更新，不是迫不得已，最好后续的更新都是兼容性的，这对使用者会很友好。

从功能上考虑，先来定义 Cascader 的 prop。

- data：决定了级联面板的内容。
- value：当前选择项，可使用 v-model。
- disabled：是否禁用。
- clearable：是否可清空。
- placeholder：占位提示。
- size：尺寸（iView 多数表单类组件都有尺寸）。
- trigger：触发方式（点击或鼠标滑入）。
- changeOnSelect：选择即改变。
- renderFormat：自定义显示内容。

对应的代码如下：

```
// cascader.vue
<script>
    export default {
        props: {
            data: {
                type: Array,
```

```js
        default () {
            return [];
        }
    },
    value: {
        type: Array,
        default () {
            return [];
        }
    },
    disabled: {
        type: Boolean,
        default: false
    },
    clearable: {
        type: Boolean,
        default: true
    },
    placeholder: {
        type: String,
        default: '请选择'
    },
    size: {
        validator (value) {
            return oneOf(value, ['small', 'large']);
        }
    },
    trigger: {
        validator (value) {
            return oneOf(value, ['click', 'hover']);
        },
        default: 'click'
    },
    changeOnSelect: {
        type: Boolean,
        default: false
    },
    renderFormat: {
        type: Function,
        default (label) {
            return label.join(' / ');
        }
    }
}
```

```
        }
    };
</script>
```

Cascader 的核心是用到了组件递归,在本书 7.5.1 小节有介绍相关用法。使用组件递归必不可少的两个条件是有 name 选项和在适当的时候结束递归。图 12-2 中所示的级联选择面板每一列都是一个组件 Caspanel（caspanel.vue），data 中的 children 决定了每项的子集,也就是需要递归显示 Caspanel 的数量。

看一下 Caspanel 的相关代码：

```
// caspanel.vue
<template>
    <span>
        <ul v-if="data && data.length" :class="[prefixCls + '-menu']">
            <Casitem
                v-for="item in data"
                :key="item"
                :prefix-cls="prefixCls"
                :data="item"
                :tmp-item="tmpItem"
                @click.native.stop="handleClickItem(item)"
                @mouseenter.native.stop="handleHoverItem(item)"></Casitem>
        </ul><Caspanel v-if="sublist && sublist.length"
:prefix-cls="prefixCls" :data="sublist" :disabled="disabled"
:trigger="trigger" :change-on-select="changeOnSelect"></Caspanel>
    </span>
</template>
<script>
    import Casitem from './casitem.vue';

    export default {
        name: 'Caspanel',
        components: { Casitem },
        props: {
            data: {
                type: Array,
                default () {
                    return [];
                }
            },
            disabled: Boolean,
            changeOnSelect: Boolean,
            trigger: String,
            prefixCls: String
```

```
        },
        data () {
            return {
                tmpItem: {},
                result: [],
                sublist: []
            };
        },
        watch: {
            data () {
                this.sublist = [];
            }
        }
    };
</script>
```

当点击某一列的某一项时，会把它对应 data 的 children 数据赋给 sublist，sublist 会作为下一个递归的 Caspanel 的 data 使用，以此类推。若该项没有 children，说明它是级联选择的最后一项，则点击直接结束选择，同时约束了 Caspanel 的递归。

最里层的组件是 Casitem，就是每列的每项，它的作用就是把 data 或 children 的每个 label 显示出来。

Cascader 的基本构成就是上述的 3 部分：cascader.vue、caspanel.vue 和 casitem.vue。cascader.vue 又分成两部分：只读输入框（Input）和下拉菜单（Drop），在下拉菜单中使用第一个 Caspanel，开始递归每一列。cascader.vue 的 template 代码为：

```
// cascader.vue
<template>
    <div :class="classes" v-clickoutside="handleClose">
        <div :class="[prefixCls + '-rel']" @click="toggleOpen">
            <slot>
                <i-input
                    readonly
                    :disabled="disabled"
                    v-model="displayRender"
                    :size="size"
                    :placeholder="placeholder"></i-input>
                <Icon
                    type="ios-close" :class="[prefixCls + '-arrow']"
                    v-show="showCloseIcon"
                    @click.native.stop="clearSelect"></Icon>
                <Icon type="arrow-down-b" :class="[prefixCls + '-arrow']">
</Icon>
            </slot>
```

```html
            </div>
            <transition name="slide-up">
                <Drop v-show="visible">
                    <div>
                        <Caspanel
                            ref="caspanel"
                            :prefix-cls="prefixCls"
                            :data="data"
                            :disabled="disabled"
                            :change-on-select="changeOnSelect"
                            :trigger="trigger"></Caspanel>
                    </div>
                </Drop>
            </transition>
        </div>
</template>
<script>
    import iInput from '../input/input.vue';
    import Drop from '../select/dropdown.vue';
    import Icon from '../icon/icon.vue';
    import Caspanel from './caspanel.vue';
    import clickoutside from '../../directives/clickoutside';

    // CSS 命名空间
    const prefixCls = 'ivu-cascader';

    export default {
        name: 'Cascader',
        components: { iInput, Drop, Icon, Caspanel },
        // 点击外部关闭的自定义指令，详见 8.2.1
        directives: { clickoutside },
        props: {
            // 省略
        },
        data () {
            return {
                prefixCls: prefixCls,
                visible: false,
                selected: [],
                tmpSelected: [],
                updatingValue: false,
                // 用于实现 v-model
                currentValue: this.value
            };
```

```
        },
        computed: {
            classes () {
                return [
                    `${prefixCls}`,
                    {
                        [`${prefixCls}-show-clear`]: this.showCloseIcon,
                        [`${prefixCls}-visible`]: this.visible,
                        [`${prefixCls}-disabled`]: this.disabled
                    }
                ];
            },
            showCloseIcon () {
                return this.currentValue && this.currentValue.length && this.clearable && !this.disabled;
            },
            // 自定义显示内容
            displayRender () {
                let label = [];
                for (let i = 0; i < this.selected.length; i++) {
                    label.push(this.selected[i].label);
                }
                return this.renderFormat(label, this.selected);
            }
        }
    };
</script>
```

Input（i-input）组件在默认的 slot 内，这意味着你可以自定义触发器部分，不局限于使用输入框，这让 Cascader 使用更灵活。使用 slot 时，需要自己渲染显示的内容，所以提供了事件 on-change，在选择完成时触发，返回 value 和 selectedData，分别为已选值和已选项的具体数据。示例代码如下：

```
<template>
    {{ text }}
    <Cascader :data="data" @on-change="handleChange">
        <a href="javascript:void(0)">选择</a>
    </Cascader>
</template>
<script>
    export default {
        data () {
            return {
                text: '未选择',
```

```
            data: [
                // 省略
            ]
        }
    },
    methods: {
        handleChange (value, selectedData) {
            this.text = selectedData.map(o => o.label).join(', ');
        }
    }
}
</script>
```

Cascader 各个组件之间的通信关系如图 12-3 所示。

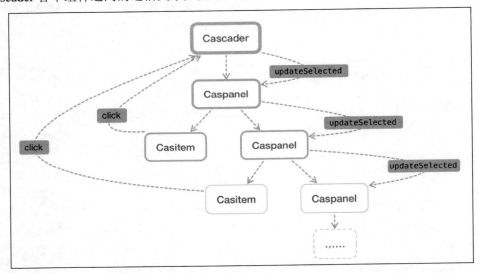

图 12-3　Cascader 组件的通信关系

我们已经多次介绍过，在 Vue 2 里，组件间通信可以通过$emit、bus、vuex 来实现。但是 iView 作为独立组件，无法使用 bus 和 vuex，为了实现跨组件通信，iView 模拟了 Vue 1 的 dispatch 和 broadcast 方法。代码如下：

```
// src/mixins/emitter.js
function broadcast(componentName, eventName, params) {
    this.$children.forEach(child => {
        const name = child.$options.name;

        if (name === componentName) {
            child.$emit.apply(child, [eventName].concat(params));
        } else {
            broadcast.apply(child, [componentName, eventName].concat([params]));
```

```
            }
        });
    }
export default {
    methods: {
        dispatch(componentName, eventName, params) {
            let parent = this.$parent || this.$root;
            let name = parent.$options.name;

            while (parent && (!name || name !== componentName)) {
                parent = parent.$parent;

                if (parent) {
                    name = parent.$options.name;
                }
            }
            if (parent) {
                parent.$emit.apply(parent, [eventName].concat(params));
            }
        },
        broadcast(componentName, eventName, params) {
            broadcast.call(this, componentName, eventName, params);
        }
    }
};
```

emitter.js 使用递归向上或向下的方式查找指定的组件名称（name），找到后触发$emit 。有了 emitter.js 就可以自由的跨组件通信了。图 12-3 中，在初始化时（mounted），Cascader 需要判断是否已经设置了选中值，若设置了，则所有的 Caspanel 和 Casitem 更新选中状态。这个过程是在 Cascader 使用 broadcast 通知 Caspanel，然后递归通知附属的 Caspanel。同理，当从父组件修改 value 时，也执行检查（即 updateSelected 方法）。在点击 Casitem 时，使用 dispatch 通知 Cascader 来更新在输入框中的选中值，这样基本就形成了一个闭环。下面来看一下与通信相关的代码：

```
// cascader.vue
<script>
    import Emitter from '../../mixins/emitter';
    export default {
        mixins: [ Emitter ],
        methods: {
            updateSelected (init = false) {
                if (!this.changeOnSelect || init) {
                    // 通知 Caspanel 更新当前选中值
                    this.broadcast('Caspanel', 'on-find-selected', {
                        value: this.currentValue
                    });
```

```js
                },
                emitValue (val, oldVal) {
                    if (JSON.stringify(val) !== oldVal) {
                        // 暴露接口
                        this.$emit('on-change', this.currentValue,
JSON.parse(JSON.stringify (this.selected)));
                    }
                }
            },
            mounted () {
                // 初始化时更新选中数据
                this.updateSelected(true);
                // 当点击 casitem 时,会派发事件到这里
                this.$on('on-result-change', (params) => {
                    const lastValue = params.lastValue;
                    const changeOnSelect = params.changeOnSelect;
                    const fromInit = params.fromInit;

                    if (lastValue || changeOnSelect) {
                        const oldVal = JSON.stringify(this.currentValue);
                        this.selected = this.tmpSelected;

                        let newVal = [];
                        this.selected.forEach((item) => {
                            newVal.push(item.value);
                        });

                        if (!fromInit) {
                            this.updatingValue = true;
                            this.currentValue = newVal;
                            this.emitValue(this.currentValue, oldVal);
                        }
                    }
                    if (lastValue && !fromInit) {
                        this.handleClose();
                    }
                });
            },
            watch: {
                // 每次展开下拉面板时都更新一次选中的数据
                visible (val) {
                    if (val) {
```

```js
            if (this.currentValue.length) {
                this.updateSelected();
            }
        }
        this.$emit('on-visible-change', val);
    },
    // v-model 的基本实现方法
    value (val) {
        this.currentValue = val;
        if (!val.length) this.selected = [];
    },
    currentValue () {
        this.$emit('input', this.currentValue);
        if (this.updatingValue) {
            this.updatingValue = false;
            return;
        }
        this.updateSelected(true);
    },
    // 如果数据源变了，也更新选中的数据
    data () {
        this.$nextTick(() => this.updateSelected());
    }
  }
};
</script>

// caspanel.vue
<script>
    import Emitter from '../../mixins/emitter';
    export default {
        mixins: [ Emitter ],
        methods: {
            // 点击选中
            handleClickItem (item) {
                if (this.trigger !== 'click' && item.children) return;
                this.handleTriggerItem(item);
            },
            // 鼠标滑过选中
            handleHoverItem (item) {
                if (this.trigger !== 'hover' || !item.children) return;
                this.handleTriggerItem(item);
            },
```

```js
handleTriggerItem (item, fromInit = false) {
    if (item.disabled) return;

    // 向上递归，设置临时选中值（并非真实选中）
    const backItem = this.getBaseItem(item);
    this.tmpItem = backItem;
    this.emitUpdate([backItem]);

    // 通知 Cascader 更新选中值
    if (item.children && item.children.length){
        this.sublist = item.children;
        this.dispatch('Cascader', 'on-result-change', {
            lastValue: false,
            changeOnSelect: this.changeOnSelect,
            fromInit: fromInit
        });
    } else {
        this.sublist = [];
        this.dispatch('Cascader', 'on-result-change', {
            lastValue: true,
            changeOnSelect: this.changeOnSelect,
            fromInit: fromInit
        });
    }
},
updateResult (item) {
    this.result = [this.tmpItem].concat(item);
    this.emitUpdate(this.result);
},
getBaseItem (item) {
    let backItem = Object.assign({}, item);
    if (backItem.children) {
        delete backItem.children;
    }

    return backItem;
},
emitUpdate (result) {
    if (this.$parent.$options.name === 'Caspanel') {
        this.$parent.updateResult(result);
    } else {
        this.$parent.$parent.updateResult(result);
    }
```

```
            }
        },
        mounted () {
            // 接收来自 Cascader 和 Caspanel 的更新选中值事件
            this.$on('on-find-selected', (params) => {
                const val = params.value;
                let value = [...val];
                for (let i = 0; i < value.length; i++) {
                    for (let j = 0; j < this.data.length; j++) {
                        if (value[i] === this.data[j].value) {
                            this.handleTriggerItem(this.data[j], true);
                            value.splice(0, 1);
                            this.$nextTick(() => {
                                // 继续向下递归更新选中状态
                                this.broadcast('Caspanel', 'on-find-selected', {
                                    value: value
                                });
                            });
                            return false;
                        }
                    }
                }
            });
        }
    };
</script>
```

独立组件与业务组件最大的不同是，业务组件往往针对数据的获取、整理、可视化，逻辑清晰简单，可以使用 vuex；而独立组件的复杂度更多集中在细节、交互、性能优化、API 设计上，对原生 JavaScript 有一定考验。在使用过程中，可能会有新功能的不断添加，也会发现隐藏的 bug，所以独立组件一开始逻辑和代码量并不复杂，多次迭代后会越来越冗长，当然功能也更丰富，使用更稳定。万事开头难，组件 API 的设计和可扩展性决定了组件迭代的复杂性。一开始不可能考虑到所有的细节，但是整体架构要清晰可扩展，否则很有可能重构。

iView 还有与 Cascader 类似思路的组件——树形控件（Tree），同样有着巧妙的设计，值得研究，源代码地址：https://github.com/iview/iview/tree/2.0/src/components/tree。

12.2　折叠面板组件 Collapse

折叠面板也是网站常用控件，可将一组内容区域展开或折叠，使页面干净整洁。iView 的折叠面板组件如图 12-4 所示。

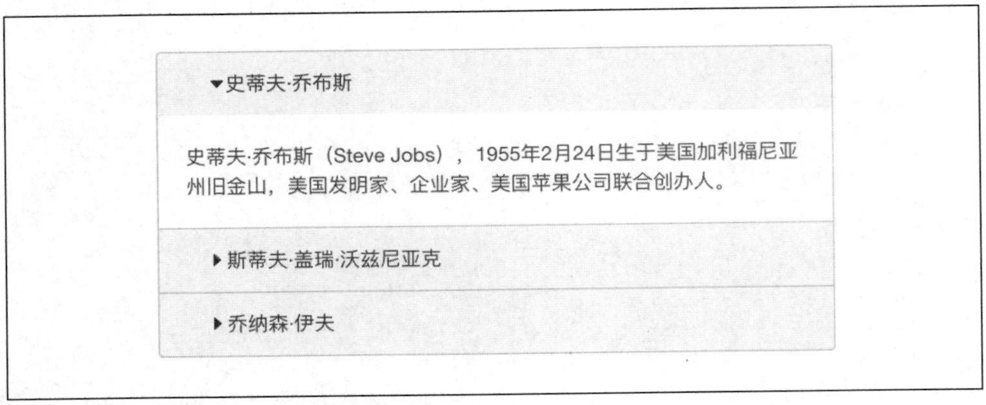

图 12-4 iView 折叠面板 Collapse

相比上一节的 Cascader，Collapse 组件无论从 UI 还是交互上都要简单得多，基于已有的知识，你完全可以自行开发出一个折叠面板组件来。通过本节的学习后，可以进一步了解组件开发中的一些细节和设计。在学习本节内容前，不妨先独立开发一个 Collapse，然后与 iView 的 Collapse 进行对比。

Collapse 对应的文档地址为 https://www.iviewui.com/components/collapse，源代码地址为 https://github.com/iview/iview/tree/2.0/src/components/collapse。

Collapse 组件分为两部分：collapse.vue 和 panel.vue，collapse 作为组件容器，接收一个整体的 slot，而 slot 就由 panel 组成，并且可以进行折叠面板的嵌套。collapse 支持 v-model 来双向绑定当前激活的面板，判断激活的依据是 panel 的 prop：name，所以 Collapse 组件的基本结构和 props 如下：

```vue
// collapse.vue
<template>
    <div :class="classes">
        <slot></slot>
    </div>
</template>
<script>
    // CSS 的命名空间
    const prefixCls = 'ivu-collapse';

    export default {
        name: 'Collapse',
        props: {
            //是否为手风琴模式，该模式下同时只能展开一个面板
            accordion: {
                type: Boolean,
                default: false
            },
            //对应panel.vue展开的name，手风琴模式下为数组
            value: {
```

```js
                    type: [Array, String]
                }
            },
            data () {
                return {
                    // 设置内部使用状态，用于实现 v-model
                    currentValue: this.value
                };
            },
            computed: {
                classes () {
                    return `${prefixCls}`;
                }
            },
            watch: {
                value (val) {
                    // 从外部改变 value 时，更新内部的数据
                    this.currentValue = val;
                }
            }
        };
</script>


// panel.vue
<template>
    <div :class="itemClasses">
        <div :class="headerClasses">
            <Icon type="arrow-right-b"></Icon>
            <slot></slot>
        </div>
        <div :class="contentClasses">
            <div :class="boxClasses">
                <slot name="content"></slot>
            </div>
        </div>
    </div>
</template>
<script>
    // 依赖 iView 的图标组件，这里使用小箭头
    import Icon from '../icon/icon.vue';
    // CSS 的命名空间
    const prefixCls = 'ivu-collapse';
```

```
    export default {
        name: 'Panel',
        components: { Icon },
        props: {
            name: {
                // 用于唯一识别当前面板
                type: String
            }
        }
    };
</script>
```

Panel 有两个 slot，默认为面板头部的内容，也就是标题，名为 content 的 slot 为主体内容。

如果没有指定 name（有些场景不关心是哪个面板，只是需要折叠面板这个功能），Collapse 就会在初始化时遍历 Panel 组件，动态地设置一个 index。Collapse 会优先识别 name，在没有定义时才使用自动设置的 index。slot：content 只在当前面板激活时显示，所以还需要增加一个数据 isActive 来控制显示与否，并通过点击默认的 slot 来切换。这部分代码如下：

```
// panel.vue，部分代码省略
<template>
    <div :class="itemClasses">
        <div :class="headerClasses" @click="toggle">
            <Icon type="arrow-right-b"></Icon>
            <slot></slot>
        </div>
        <div :class="contentClasses" v-show="isActive">
            <div :class="boxClasses">
                <slot name="content"></slot>
            </div>
        </div>
    </div>
</template>
<script>
    export default {
        data () {
            return {
                index: 0,
                isActive: false
            };
        },
        methods: {
            toggle () {
```

```js
            // 访问父链（即 collapse.vue）执行方法，稍后介绍
            this.$parent.toggle({
                // 优先使用 name，未定义时使用 index
                // index 和 isActive 都在 collapse.vue 中设置，稍后介绍
                name: this.name || this.index,
                isActive: this.isActive
            });
        }
    },
    // 动态设置相关 CSS 类名
    computed: {
        itemClasses () {
            return [
                `${prefixCls}-item`,
                {
                    [`${prefixCls}-item-active`]: this.isActive
                }
            ];
        },
        headerClasses () {
            return `${prefixCls}-header`;
        },
        contentClasses () {
            return `${prefixCls}-content`;
        },
        boxClasses () {
            return `${prefixCls}-content-box`;
        }
    }
};
</script>
```

以上是所有 Panel 组件的代码。接下来在 collapse.vue 初始化时，通过访问子链（即遍历所有的 panel.vue）来给 panel 动态设置 index 和 isActive。Collapse 中有两个方法：getActiveKey 和 setActive，前者用于将当前激活面板的受控数据 currentValue 根据是否为手风琴状态做处理，使两种模式下的数据格式统一。setActive 是遍历 Panel，并设置其 index 和 isActive。相关代码如下：

```js
// collapse.vue，部分代码省略
export default {
    methods: {
        getActiveKey () {
            let activeKey = this.currentValue || [];
            const accordion = this.accordion;
```

```js
            /*
             * value 的类型可以是字符串或数组，
             * 为保证数据格式统一，将 activeKey 强行设置为数组因为不能保证用户设置的都是数组
             */
            if (!Array.isArray(activeKey)) {
                activeKey = [activeKey];
            }
            // 手风琴模式下，如果多设置了数组，也只取其第一项
            if (accordion && activeKey.length > 1) {
                activeKey = [activeKey[0]];
            }
            //将 activeKey 转为字符串，因为用户设置的有可能是字符串数字，比较时类型会不一致
            for (let i = 0; i < activeKey.length; i++) {
                activeKey[i] = activeKey[i].toString();
            }

            return activeKey;
        },
        setActive () {
            const activeKey = this.getActiveKey();
            this.$children.forEach((child, index) => {
                const name = child.name || index.toString();
                let isActive = false;
                //在两种模式下，判断当前 Panel 是否需要激活
                if (self.accordion) {
                    isActive = activeKey === name;
                } else {
                    isActive = activeKey.indexOf(name) > -1;
                }
                //给当前 Panel 设置 isActive 和 index，index 为它在 slot 中的序列
                child.isActive = isActive;
                child.index = index;
            });
        }
    },
    mounted () {
        // 初始化时，设置 isActive 和 index
        this.setActive();
    },
    watch: {
        value (val) {
            this.currentValue = val;
        },
```

```
            //修改 currentValue 时，重新设置一遍状态
            currentValue () {
                this.setActive();
            }
        }
    };
```

最后还剩余在 Panel 中遗留的一个 toggle 方法，当切换面板的激活与隐藏状态时，需要更新 currentValue，并发出自定义事件 input 和 on-change，其中 input 是为了实现 v-model 语法糖，代码如下：

```
// collapse.vue，部分代码省略
export default {
    methods: {
        /*
         * data 由 Panel 传递，有两项：
         * name，当前 Panel 的 name 或 index 的值
         * isActive，当前是否激活
         */
        toggle (data) {
            const name = data.name.toString();
            // 声明一个临时激活项列表
            let newActiveKey = [];
            if (this.accordion) {
                /*
                 * 手风琴模式下，同时只能激活一个面板
                 * 如果当前未激活，就将它的 name 写入临时列表
                 * 如果已激活，意味着关闭所有的面板，所以不用任何设置
                 */
                if (!data.isActive) {
                    newActiveKey.push(name);
                }
            } else {
                let activeKey = this.getActiveKey();
                const nameIndex = activeKey.indexOf(name);
                // 点击后切换为关闭
                if (data.isActive) {
                    if (nameIndex > -1) {
                        activeKey.splice(nameIndex, 1);
                    }
                // 点击后切换为展开
                } else {
                    if (nameIndex < 0) {
```

```
                    activeKey.push(name);
                }
            }
            newActiveKey = activeKey;
        }
        // 更新 currentValue 会触发 watch，从而调用 setActive 方法
        this.currentValue = newActiveKey;
        this.$emit('input', newActiveKey);
        this.$emit('on-change', newActiveKey);
    }
}
```

以上就是 Collapse 组件所有的代码，下面为图 12-4 对应的代码：

```
<template>
    <Collapse v-model="value">
        <Panel name="1">
            史蒂夫·乔布斯
            <p slot="content">史蒂夫·乔布斯（Steve Jobs）……</p>
        </Panel>
        <Panel name="2">
            斯蒂夫·盖瑞·沃兹尼亚克
            <p slot="content">斯蒂夫·盖瑞·沃兹尼亚克（Stephen Gary Wozniak）……</p>
        </Panel>
        <Panel name="3">
            乔纳森·伊夫
            <p slot="content">乔纳森·伊夫……</p>
        </Panel>
    </Collapse>
</template>
<script>
    export default {
        data () {
            return {
                value: '1'
            }
        }
    }
</script>
```

在 Panel 中还可以嵌套 Collapse，如图 12-5 所示。

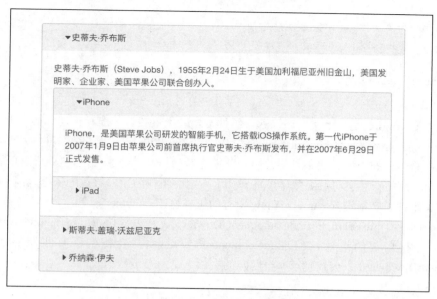

图 12-5　Collapse 的嵌套用法

iView 的 40 多个组件都是独立的 UI 组件，它无法像业务组件那样使用 Vuex、bus 等技术进行跨组件通信，因此会经常访问和操作父（子）链来修改状态及调用方法。但是在业务开发中，要尽量避免这样的操作，因为很难知道是谁修改了组件的状态，正确的做法应该是使用 Vuex 或 bus 来统一维护。

12.3　iView 内置工具函数

iView 项目中还有很多实用的工具函数，在 https://github.com/iview/iview/blob/2.0/src/utils/assist.js 文件中。比如 findComponentUpward、findComponentDownward 和 findComponentsDownward 方法，它们用来向上或向下寻找指定 name 的组件，有些场景下会比上一节介绍的 broadcast 和 dispatch 方法好用，因为这 3 个方法直接返回的是组件实例，而不是传递数据。

findComponentUpward 方法以当前实例为参照点，向上寻找出指定 name 或几个 name 中的一个组件实例，找到后立即返回该实例。函数代码如下：

```
function findComponentUpward (context, componentName, componentNames) {
    if (typeof componentName === 'string') {
        componentNames = [componentName];
    } else {
        componentNames = componentName;
    }

    let parent = context.$parent;
    let name = parent.$options.name;
    while (parent && (!name || componentNames.indexOf(name) < 0)) {
```

```
            parent = parent.$parent;
            if (parent) name = parent.$options.name;
        }
        return parent;
    }
```

第一个参数 context 是上下文，即以哪个组件开始向上寻找，一般都传递 this，也就是当前的实例。componentName 和 componentNames 只需要传递一个即可，前者是字符串，后者是数组，函数开始会判断传递的类型，如果是字符串，就把它转为一个数组来使用，保证格式统一。

使用 while 语句一层层向上级循环，直到找出指定的组件为止，寻找的依据是组件的 name 字段，所以在写组件时必须设置 name，iView 所有的组件也都有 name 字段。

findComponentDownward 和 findComponentsDownward 方法与 findComponentUpward 类似，不同的是向下寻找指定的组件，但 findComponentsDownward 会找到所有匹配的子组件，而 findComponentDownward 只会找到第一个匹配的。以下是两个函数的代码：

```
// Find component downward
function findComponentDownward (context, componentName) {
    const childrens = context.$children;
    let children = null;
    if (childrens.length) {
        childrens.forEach(child => {
            const name = child.$options.name;
            if (name === componentName) {
                children = child;
            }
        });
        for (let i = 0; i < childrens.length; i++) {
            const child = childrens[i];
            const name = child.$options.name;
            if (name === componentName) {
                children = child;
                break;
            } else {
                children = findComponentDownward(child, componentName);
                if (children) break;
            }
        }
    }
    return children;
}
// Find components downward
function findComponentsDownward (context, componentName, components = []) {
    const childrens = context.$children;
```

```
        if (childrens.length) {
            childrens.forEach(child => {
                const name = child.$options.name;
                const childs = child.$children;
                if (name === componentName) components.push(child);
                if (childs.length) {
                    const findChilds = findComponentsDownward(child, componentName, components);
                    if (findChilds) components.concat(findChilds);
                }
            });
        }
        return components;
    }
```

iView 在 Radio、Checkbox、Menu 等很多组件中都使用了这 3 个方法。如果你已经使用了 iView 项目，那么 assist.js 里所有的方法都可以使用。以这 3 个方法为例，在业务中可以这样导入：

```
import {
    findComponentUpward,
    findComponentDownward,
    findComponentsDownward
} from 'iview/src/utils/assist';
```

除了工具函数外，iView 内置的自定义指令、混合也可以直接使用。自定义指令所在地址为 https://github.com/iview/iview/tree/2.0/src/directives，其中 clickoutside.js 已经在第 8 章中有所介绍，是用来点击外部关闭弹窗用的。transfer-dom.js 用于将当前 dom 插入 body 内，iView 的 Modal 组件中使用过，模态框（Modal）弹出时，会覆盖整个屏幕，如图 12-6 所示。

图 12-6　iView 的 Modal 组件

如果<Modal></Modal>所在的 DOM 使用了 CSS 定位，那么 Modal 的 fixed 定位参照会不准确，因为它相对的是当前实例所在的 DOM，使用 transfer-dom 指令后，Modal 会被移动到 body 内，也就是相对整个 body 使用 fixed 定位，这样避免了遮罩不完全的情况。

iView 同时也是一整套的前端解决方案，包含工程构建、主题定制、多语言等功能，极大地提升了开发效率。如果你的项目是面向中后台业务的，不妨试试这套 UI 框架，相信至少会提高一倍的开发效率。

第 13 章

实战：知乎日报项目开发

知乎日报是由知乎开发的一款资讯类阅读 App，每日提供来自知乎社区精选的问答或专栏文章。本章将使用 Vue 和 webpack 等相关技术，利用知乎日报的接口开发一个 Web App。

13.1 分析与准备

本章将以第 10 章的 webpack 配置为基础进行开发，可以先到 GitHub 下载工程配置文件：https://github.com/icarusion/vue-book/tree/master/demo ，将项目保存到新建的 daily 目录，然后完成依赖安装。本章所有的代码也上传至 GitHub，访问链接可以查看并直接使用：https://github.com/icarusion/vue-book/tree/master/daily 。

日报是一个单页的应用，由 3 部分组成，如图 13-1 所示。

左侧是菜单，分为"每日推荐"和"主题日报"两个类型，中间是文章列表，右侧是文章正文和评论。其中每日推荐按日期排列，比如图中显示为 5 月 2 日的推荐文章，中间栏滚动至底部时，自动加载前一天的推荐内容。

主题日报有"日常心理学"等 10 多个子分类，分类列表默认是收起的，点击"主题日报"菜单时切换展开和收起的状态。点击某个子分类后，中间栏切换为该类目下的文章列表，不再按时间排列。点击文章列表中的某一项，在右侧渲染对应文章的内容和评论。

知乎日报的接口地址前缀为 http://news-at.zhihu.com/api/4/ ，图片地址前缀为 https://pic1.zhimg.com，由于两者都开启了跨域限制，无法在前端直接调用，因此要开发一个代理。

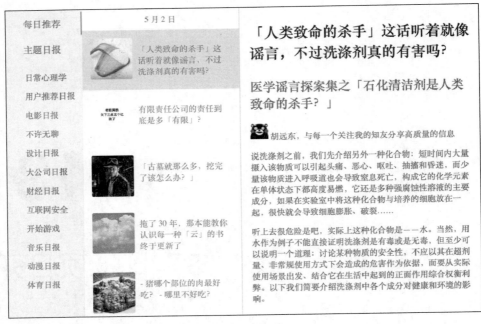

图 13-1　日报应用效果图

提　示

跨域限制是服务端的一个行为，当开启对某些域名的访问限制后，只有同域或指定域下的页面可以调用，这样相对来说更安全，图片也可以防盗链。跨域限制一般只在浏览器端存在，对于服务端或 iOS、Android 等客户端是不存在的。使用代理是开发过程中常见的一种解决方案。

我们使用基于 Node.js 的 request 库来做代理，通过 NPM 安装 request：

```
npm install request --save-dev
```

在 daily 目录下新建一个 proxy.js 的文件，写入以下内容（如果你不太了解 Node.js，可以先直接使用）：

```javascript
const http = require('http');
const request = require('request');

const hostname = '127.0.0.1';
const port = 8010;
const imgPort = 8011;

// 创建一个 API 代理服务
const apiServer = http.createServer((req, res) => {
    const url = 'http://news-at.zhihu.com/api/4' + req.url;
    const options = {
        url: url
    };
```

```javascript
    function callback (error, response, body) {
        if (!error && response.statusCode === 200) {
            // 设置编码类型，否则中文会显示为乱码
            res.setHeader('Content-Type', 'text/plain;charset=UTF-8');
            // 设置所有域允许跨域
            res.setHeader('Access-Control-Allow-Origin', '*');
            //返回代理后的内容
            res.end(body);
        }
    }
    request.get(options, callback);
});
// 监听 8010 端口
apiServer.listen(port, hostname, () => {
    console.log(`接口代理运行在 http://${hostname}:${port}/`);
});
// 创建一个图片代理服务
const imgServer = http.createServer((req, res) => {
    const url = req.url.split('/img/')[1];
    const options = {
        url: url,
        encoding: null
    };

    function callback (error, response, body) {
        if (!error && response.statusCode === 200) {
            const contentType = response.headers['content-type'];
            res.setHeader('Content-Type', contentType);
            res.setHeader('Access-Control-Allow-Origin', '*');
            res.end(body);
        }
    }
    request.get(options, callback);
});
// 监听 8011 端口
imgServer.listen(imgPort, hostname, () => {
    console.log(`图片代理运行在 http://${hostname}:${imgPort}/`);
});
```

监听了两个端口：8010 和 8011。8010 用于接口代理，8011 用于图片代理。比如请求的真实接口为 http://news-at.zhihu.com/api/4/news/3892357，开发时改写为 http://127.0.0.1:8010/news/3892357；图片的真实地址为 https://pic4.zhimg.com/v2-b44636ccd2affac97ccc0759a0f46f7f.jpg，开发

时改写为 http://127.0.0.1:8011/img/https://pic4.zhimg.com/v2-b44636ccd2affac97ccc0759a0f46f7f.jpg。

代理的核心是在返回的头部（response header）中添加一项 Access-Control-Allow-Origin 为"*"，也就是允许所有的域访问。

最后在终端使用 Node 启动代理服务：

node proxy.js

如果成功，就会在终端显示两行日志：

接口代理运行在 http://127.0.0.1:8010/

图片代理运行在 http://127.0.0.1:8011/

对于接口的 Ajax 请求，前端有很多实现方案，比如 jQuery 的$.ajax，但是只为使用 Ajax 而引入一个 jQuery 显然不太友好。如果你完成了 11.3 章节的练习题，可以用做好的 Vue 插件通过 this.$ajax 直接请求。Vue 官方也提供了 vue-resource 插件，但是不再维护，而是推荐使用 axios，所以本章示例也基于 axios 来做异步请求。

axios 是基于 Promise 的 HTTP 库，同时支持前端和 Node.js。首先用 NPM 安装 axios：

npm install axios --save

在 daily 目录下新建目录 libs，并在 libs 下新建 util.js 文件，项目中使用的工具函数可以在这里封装。比如对 axios 封装，写入请求地址的前缀，在业务中只用写相对路径，这样可以灵活控制。另外，可以全局拦截 axios 返回的内容，简单处理，只需返回我们需要的数据。其代码如下：

```
// util.js
import axios from 'axios';
// 基本配置
const Util = {
    imgPath: 'http://127.0.0.1:8011/img/',
    apiPath: 'http://127.0.0.1:8010/'
};
// Ajax 通用配置
Util.ajax = axios.create({
    baseURL: Util.apiPath
});
//添加响应拦截器
Util.ajax.interceptors.response.use(res => {
    return res.data;
});

export default Util;
```

更多关于 axios 的使用可以查阅官方文档：https://github.com/mzabriskie/axios。

做好这些准备后，就可以开始日报应用的开发了。

13.2 推荐列表与分类

13.2.1 搭建基本结构

项目中使用的 CSS 样式不多，所以直接写在 daily/style.css，并在 main.js 中导入：

```
// main.js
import Vue from 'vue';
import App from './app.vue';
import './style.css';

new Vue({
    el: '#app',
    render: h => {
        return h(App)
    }
});
```

日报是单页应用，没有路由，只有一个入口组件 app.vue。应用结构如图 13-2 所示。

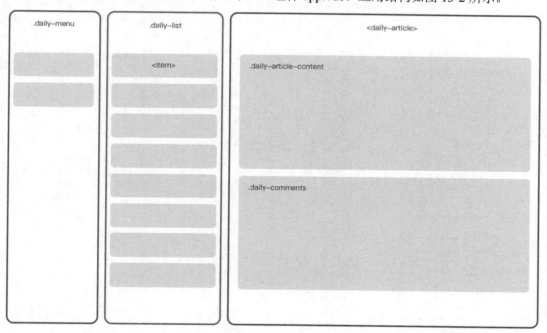

图 13-2　日报应用结构

应用分左、中、右 3 栏，3 栏都可以滚动。对左栏和中栏使用 fixed 固定，并使用 overflow: auto 滚动，而右栏高度自适应，使用浏览器默认的 body 区域滚动即可。基本的 HTML 和 CSS 结构如下：

```
// app.vue
<template>
    <div class="daily">
        <div class="daily-menu">
            <div class="daily-menu-item">每日推荐</div>
            <div class="daily-menu-item">主题日报</div>
        </div>
        <div class="daily-list">
            <Item></Item>
        </div>
        <daily-article></daily-article>
    </div>
</template>

// style.css
html, body{
    margin: 0;
    padding: 0;
    height: 100%;
    color: #657180;
    font-size: 16px;
}
.daily-menu{
    width: 150px;
    position: fixed;
    top: 0;
    bottom: 0;
    left: 0;
    overflow: auto;
    background: #f5f7f9;
}
.daily-menu-item{
    font-size: 18px;
    text-align: center;
    margin: 5px 0;
    padding: 10px 0;
    cursor: pointer;
    border-right: 2px solid transparent;
    transition: all .3s ease-in-out;
}
.daily-menu-item:hover{
    background: #e3e8ee;
}
```

```css
.daily-menu-item.on{
    border-right: 2px solid #3399ff;
}

.daily-list{
    width: 300px;
    position: fixed;
    top: 0;
    bottom: 0;
    left: 150px;
    overflow: auto;
    border-right: 1px solid #d7dde4;
}
.daily-item{
    display: block;
    color: inherit;
    text-decoration: none;
    padding: 16px;
    overflow: hidden;
    cursor: pointer;
    transition: all .3s ease-in-out;
}
.daily-item:hover{
    background: #e3e8ee;
}
.daily-article{
    margin-left: 450px;
    padding: 20px;
}
```

13.2.2 主题日报

"主题日报"下有子类列表,默认是收起的,点击主题日报可以切换展开和收起的状态,使用数据 showThemes 来控制,并用 themes 来循环渲染子类目:

```
// app.vue,部分代码省略
<template>
    <div class="daily-menu">
        <div class="daily-menu-item"
            :class="{ on: type === 'recommend' }">每日推荐</div>
        <div class="daily-menu-item"
            :class="{ on: type === 'daily' }"
            @click="showThemes = !showThemes">主题日报</div>
        <ul v-show="showThemes">
```

```html
            <li v-for="item in themes">
                <a :class="{ on: item.id === themeId && type === 'daily' }">{{ item.name }}</a>
            </li>
        </ul>
    </div>
</template>
<script>
    export default {
        data () {
            return {
                themes: [],
                showThemes: false,
                type: 'recommend',
                themeId: 0
            }
        },
    }
</script>
// style.css
.daily-menu ul{
    list-style: none;
}
.daily-menu ul li a{
    display: block;
    color: inherit;
    text-decoration: none;
    padding: 5px 0;
    margin: 5px 0;
    cursor: pointer;
}
.daily-menu ul li a:hover, .daily-menu ul li a.on{
    color: #3399ff;
}
```

themeId 会在点击子类时设置，稍后会介绍。

应用初始化时，获取主题日报的分类列表：

```html
// app.vue，部分代码省略
<script>
    import $ from './libs/util';
    export default {
        data () {
            return {
                themes: []
```

```
            }
        },
        methods: {
            getThemes () {
                // axios 发起 get 请求
                $.ajax.get('themes').then(res => {
                    this.themes = res.others;
                })
            }
        },
        mounted () {
            // 初始化时调用
            this.getThemes();
        }
    }
</script>
```

主题日报类目列表为数组，每一项的结构示例如下：

```
"others": [
    {
        "name": "日常心理学",
        "id": 13,
        "thumbnail": "http://pic3.zhimg.com/xxx.jpg",
        "color": 15007,
        "description": "了解自己和别人，了解彼此的欲望和局限。"
    }
]
```

点击子类目时，将菜单 type 切换为"主题日报"高亮点击的子类，然后加载该类目下的文章列表：

```
// app.vue，部分代码省略
<template>
    <ul v-show="showThemes">
        <li v-for="item in themes">
            <a
                :class="{ on: item.id === themeId && type === 'daily' }"
                @click="handleToTheme(item.id)">{{ item.name }}</a>
        </li>
    </ul>
</template>
<script>
    import $ from './libs/util';
    export default {
        data () {
```

```
        return {
            themes: [],
            showThemes: false,
            type: 'recommend',
            list: [],
            themeId: 0
        }
    },
    methods: {
        handleToTheme (id) {
            // 改变菜单分类
            this.type = 'daily';
            // 设置当前点击子类的主题日报 id
            this.themeId = id;
            // 清空中间栏的数据
            this.list = [];
            $.ajax.get('theme/' + id).then(res => {
                // 过滤掉类型为 1 的文章，该类型下的文章为空
                this.list = res.stories
                    .filter(item => item.type !== 1);
            })
        }
    }
}
</script>
```

文章列表 list 为数组，每一项的结构示例如下：

```
"stories": [
    {
        "type": 0,
        "id": 7097426,
        "title": "人们在虚拟生活中投入的精力是否对现实生活的人际关系有积极意义？"
    },
    {
        "type": 0,
        "id": 7101963,
        "title": "写给想成为心理咨询师的学生同仁",
        "images": [
            "http://pic1.zhimg.com/xxx.jpg"
        ]
    }
]
```

文章列表中的 id 字段是文章的 id，请求文章内容和评论列表时会用到，title 为标题，images 为封面图片，没有 images 字段就不显示封面图片。

13.2.3 每日推荐

应用初始化和点击"每日推荐"菜单时请求推荐的文章列表。推荐列表的 API 相对地址为 news/before/20170503，before 后面是查询的日期，这个日期比要查询的真实日期多一天，比如要查 2017 年 5 月 2 日推荐的内容，就要请求 20170503。每日推荐可以无限次地向前一天查询，为方便操作日期，在 libs/util.js 内定义两个时间方法：

```javascript
// libs/util.js，部分代码省略
import axios from 'axios';
const Util = {

};

// 获取今天的时间戳
Util.getTodayTime = function () {
    const date = new Date();
    date.setHours(0);
    date.setMinutes(0);
    date.setSeconds(0);
    date.setMilliseconds(0);
    return date.getTime();
};
// 获取前一天的日期
Util.prevDay = function (timestamp = (new Date()).getTime()) {
    const date = new Date(timestamp);
    const year = date.getFullYear();
    const month = date.getMonth() + 1 < 10
        ? '0' + (date.getMonth() + 1)
        : date.getMonth() + 1;
    const day = date.getDate() < 10
        ? '0' + date.getDate()
        : date.getDate();
    return year + '' + month + '' + day;
};
export default Util;
```

Util.prevDay 的参数为前一天的时间戳，计算前一天的时间戳只需以今天 0 点的时间戳为基础，也就是通过 Util.getTodayTime 获取的时间戳减去 86400000（24*60*60*1000）。这种方法要比直接判断前一天的日期简单得多，因为每个月的日期是不固定的，另外还需特殊处理润年。

推荐文章的列表获取的相关代码如下：

```
// app.vue,部分代码省略
<template>
    <div class="daily-menu-item"
        @click="handleToRecommend"
        :class="{ on: type === 'recommend' }">每日推荐</div>
</template>
<script>
    import $ from './libs/util';
    export default {
        data () {
            return {
                type: 'recommend',
                recommendList: [],
                dailyTime: $.getTodayTime(),
                isLoading: false
            }
        },
        methods: {
            handleToRecommend () {
                this.type = 'recommend';
                this.recommendList = [];
                this.dailyTime = $.getTodayTime();
                this.getRecommendList();
            },
            getRecommendList () {
                this.isLoading = true;
                const prevDay = $.prevDay(this.dailyTime + 86400000);
                $.ajax.get('news/before/' + prevDay).then(res => {
                    this.recommendList.push(res);
                    this.isLoading = false;
                })
            }
        },
        mounted () {
            this.getRecommendList();
        }
    }
</script>
```

recommendList 为推荐文章列表的数据，在初始化和每次点击"每日推荐"菜单时都会请求数据。dailyTime 默认获取今天 0 点的时间戳，请求时需要多加一天。因为推荐列表可能通过"主题日报"的子类切换而来，需要重新获取一遍数据，所以 handleToRecommend 方法每次都需要清空列表并重新设置 dailyTime。

推荐列表的数据结构和主题日报基本一致，不同的是多了一个 date 字段来表示请求列表的日期，比如：

```
{
    "date": "20170502",
    "stories": [
        {
            "id": 9394848,
            "title": "在庞大的体系中像齿轮一样工作，如何避免"去能力化"？",
            "images": [
                "https://pic4.zhimg.com/xxx.jpg"
            ],
            "ga_prefix": "050220",
            "type": 0
        }
    ]
}
```

两个文章列表（list、recommendList）的每项都用一个组件 item.vue 来展示，在 daily/components 目录下新建 item.vue 文件，并写入以下内容：

```
// components/item.vue
<template>
    <a class="daily-item">
        <div class="daily-img" v-if="data.images">
            <img :src="imgPath + data.images[0]">
        </div>
        <div
            class="daily-title"
            :class="{ noImg: !data.images}">{{ data.title }}</div>
    </a>
</template>
<script>
    import $ from '../libs/util';
    export default {
        props: {
            data: {
                type: Object
            }
        },
        data () {
            return {
                imgPath: $.imgPath
            }
```

```
        }
    }
</script>
// style.css
.daily-item{
    display: block;
    color: inherit;
    text-decoration: none;
    padding: 16px;
    overflow: hidden;
    cursor: pointer;
    transition: all .3s ease-in-out;
}
.daily-item:hover{
    background: #e3e8ee;
}
.daily-img{
    width: 80px;
    height: 80px;
    float: left;
}
.daily-img img{
    width: 100%;
    height: 100%;
    border-radius: 3px;
}
.daily-title{
    padding: 10px 5px 10px 90px;
}
.daily-title.noImg{
    padding-left: 5px;
}
```

prop：data 里可能没有 images 字段，所以列表会显示两种模式，即含封面图和不含封面图。

Item 组件会用到文章列表里，type 为 recommend 和 daily 两种类型下，渲染会稍有不同。recommend 会显示每天的日期，daily 则没有，两者效果如图 13-3 所示。

图 13-3　两种类型的文章列表对比

对应的代码如下：

```
// app.vue，部分代码省略
<template>
    <div class="daily-list">
        <template v-if="type === 'recommend'">
            <div v-for="list in recommendList">
                <div class="daily-date">{{ formatDay(list.date) }}</div>
                <Item
                    v-for="item in list.stories"
                    :data="item"
                    :key="item.id"></Item>
            </div>
        </template>
        <template v-if="type === 'daily'">
            <Item
                v-for="item in list"
                :data="item"
                :key="item.id"></Item>
        </template>
    </div>
</template>
<script>
    import Item from './components/item.vue';
```

```js
    export default {
        components: { Item },
        data () {
            return {
                type: 'recommend',
                recommendList: [],
                list: []
            }
        },
        methods: {
            // 转换为带汉字的月日
            formatDay (date) {
                let month = date.substr(4, 2);
                let day = date.substr(6, 2);
                if (month.substr(0, 1) === '0') month = month.substr(1, 1);
                if (day.substr(0, 1) === '0') day = day.substr(1, 1);
                return `${month} 月 ${day} 日`;
            }
        }
    }
</script>
// style.css
.daily-list{
    width: 300px;
    position: fixed;
    top: 0;
    bottom: 0;
    left: 150px;
    overflow: auto;
    border-right: 1px solid #d7dde4;
}
.daily-date{
    text-align: center;
    margin: 10px 0;
}
```

13.2.4 自动加载更多推荐列表

在"每日推荐"类型下,中栏的文章列表滚动到底部会自动加载前一天的推荐列表,所以要监听中栏(.daily-list)的滚动事件,并在合适的时机触发加载请求:

```html
// app.vue,部分代码省略
<template>
    <div class="daily-list" ref="list">
```

```
        </div>
</template>
<script>
    export default {
        data () {
            return {
                isLoading: false
            }
        },
        methods: {
            getRecommendList () {
                // 加载时设置为true，加载完成后置为false
                this.isLoading = true;
                const prevDay = $.prevDay(this.dailyTime + 86400000);
                $.ajax.get('news/before/' + prevDay).then(res => {
                    this.recommendList.push(res);
                    this.isLoading = false;
                })
            }
        },
        mounted () {
            this.getRecommendList();
            // 获取到DOM
            const $list = this.$refs.list;
            // 监听中栏的滚动事件
            $list.addEventListener('scroll', () => {
                //在"主题日报"或正在加载推荐列表时停止操作
                if (this.type === 'daily' || this.isLoading) return;
                // 已经滚动的距离加页面的高度等于整个内容区域高度时，视为接触底部
                if
                (
                    $list.scrollTop
                    + document.body.clientHeight
                    >= $list.scrollHeight
                )
                {
                    //时间相对减少一天
                    this.dailyTime -= 86400000;
                    this.getRecommendList();
                }
            });
        }
```

```
}
</script>
```

$list（.daily-list）的 CSS 使用了 overflow: auto，所以它具备滚动的能力，进而可以监听滚动事件。直接操作 DOM 在 Vue 中很少见，但示例的场景和一些对 window、document 对象监听事件的场景还是有的，使用监听时要注意在 beforeDestroy 生命周期使用 removeEventListener 移除，本例是单页应用，所以不需此操作。$list 的 scroll 是标准 DOM 事件，所以也可以用 Vue 的 v-on 指令，比如上例也可以改写为：

```
<template>
    <div
        class="daily-list"
        ref="list"
        @scroll="handleScroll"></div>
</template>
<script>
    export default {
        methods: {
            handleScroll () {
                const $list = this.$refs.list;
                if (this.type === 'daily' || this.isLoading) return;
                if
                (
                    $list.scrollTop
                    + document.body.clientHeight
                    >= $list.scrollHeight
                )
                {
                    this.dailyTime -= 86400000;
                    this.getRecommendList();
                }
            }
        }
    }
</script>
```

13.3　文章详情页

13.3.1　加载内容

右侧的文章内容区域封装成了一个组件。在 components 目录下新建 daily-article.vue 组件，它接收唯一的一个 prop：id，也就是文章的 id，如果 id 变化了，就说明切换了文章，需要请求新的文章内容。

在 app.vue 中导入 daily-article.vue 组件，并在文章列表的 Item 组件上绑定查看文章事件：

```
// app.vue，部分代码省略和简写
<template>
    <div class="daily">
        <Item @click.native="handleClick(item.id)"></Item>
        <daily-article :id="articleId"></daily-article>
    </div>
</template>
<script>
    import Item from './components/item.vue';
    import dailyArticle from './components/daily-article.vue';

    export default {
        components: { Item, dailyArticle },
        data () {
            return {
                articleId: 0
            }
        },
        methods: {
            handleClick (id) {
                this.articleId = id;
            }
        }
    }
</script>
```

Item 是组件，绑定原生事件时要带事件修饰符 .native，否则会认为监听的是来自 Item 组件的自定义事件 click。

dailyArticle 组件在监听到 id 改变时请求文章内容：

```
// components/daily-article.vue
<template>
    <div class="daily-article">
        <div class="daily-article-title">{{ data.title }}</div>
        <div class="daily-article-content" v-html="data.body"></div>
    </div>
</template>
<script>
    import $ from '../libs/util';
    export default {
        props: {
            id: {
```

```js
                type: Number,
                default: 0
            }
        },
        data () {
            return {
                data: {}
            }
        },
        methods: {
            getArticle () {
                $.ajax.get('news/' + this.id).then(res => {
                    //将文章的中的图片地址替换为代理的地址
                    res.body = res.body
                        .replace(/src="http/g, 'src="' + $.imgPath + 'http');
                    res.body = res.body
                        .replace(/src="https/g, 'src="' + $.imgPath + 'https');
                    this.data = res;
                    //返回文章顶端
                    window.scrollTo(0, 0);
                })
            }
        },
        watch: {
            id (val) {
                if (val) this.getArticle();
            }
        }
    };
</script>
// style.css
.daily-article{
    margin-left: 450px;
    padding: 20px;
}
.daily-article-title{
    font-size: 28px;
    font-weight: bold;
    color: #222;
    padding: 10px 0;
}
.view-more a{
    display: block;
```

```css
    cursor: pointer;
    background: #f5f7f9;
    text-align: center;
    color: inherit;
    text-decoration: none;
    padding: 4px 0;
    border-radius: 3px;
}
```

数据的 data 结构为：

```json
{
    "title": "这茶，明显是用了梅雨期的雨水，我还是喜欢用腊月的雪水",
    "body": "文章内容，格式为 html",
    "id": 9395306,
    "type": 0,
    "image": "https://pic3.zhimg.com/v2-dbf5d6e5eeeccaacc67af4d625e0699a.jpg",
    "image_source": "T.Tseng / CC BY",
    "images": [
        "https://pic4.zhimg.com/v2-5cb4fcbd56bb6717969e9967829929b7.jpg"
    ],
    "share_url": "http://daily.zhihu.com/story/9395306",
    "ga_prefix": "050311",
    "js": [],
    "css": [
        "http://news-at.zhihu.com/css/news_qa.auto.css?v=4b3e3"
    ]
}
```

这里只用到了 title 和 body，其中 body 的格式为 html，需要用 v-html 指令直接显示。用户可能会在某篇文章阅读到一定位置时切换了别的文章，这时文章的滚动条仍停留在上次浏览的位置，使用 window.scrollTo(0, 0) 可以返回页面的顶端。需要注意的是，.daily-article 并没有使用 overflow: auto 滚动，而是自然高度，所以这里是让页面返回顶端，而不能设置 .daily-article 的 scrollTop 为 0。

13.3.2 加载评论

每篇文章底部要加载评论，效果如图 13-4 所示。

评论的数据结构为：

```json
"comments": [
    {
        "author": "滕正云",
        "content": "善泳者溺于水 佩服于极限运动者的勇气 但我想这应该是小圈子内的英雄",
        "avatar": "http://pic1.zhimg.com/xxx.jpg",
```

```
        "time": 1493788345,
        "id": 28885287,
        "likes": 0
    }
]
```

图 13-4　评论列表

每条评论要显示发表时间，源数据格式为时间戳，需要前端转为相对时间。在第 8 章自定义指令的 8.2.2 小节实现了一个 v-time 的自定义指令，用于时间戳转换，可以直接使用，但是要修改为 ES6 Module 语法导出模块。在 daily 目录下创建 directives 目录，并创建 time.js 文件，写入以下内容：

```
// directives/time.js
var Time = {
    // 获取当前时间戳
    getUnix: function () {
        var date = new Date();
        return date.getTime();
    },
    // 获取今天 0 点 0 分 0 秒的时间戳
    getTodayUnix: function () {
        var date = new Date();
        date.setHours(0);
        date.setMinutes(0);
```

```javascript
        date.setSeconds(0);
        date.setMilliseconds(0);
        return date.getTime();
    },
    // 获取今年1月1日0点0分0秒的时间戳
    getYearUnix: function () {
        var date = new Date();
        date.setMonth(0);
        date.setDate(1);
        date.setHours(0);
        date.setMinutes(0);
        date.setSeconds(0);
        date.setMilliseconds(0);
        return date.getTime();
    },
    // 获取标准年月日
    getLastDate: function(time) {
        var date = new Date(time);
        var month = date.getMonth() + 1 < 10 ? '0' + (date.getMonth() + 1) : date.getMonth() + 1;
        var day = date.getDate() < 10 ? '0' + date.getDate() : date.getDate();
        return date.getFullYear() + '-' + month + "-" + day;
    },
    // 转换时间
    getFormatTime: function(timestamp) {
        var now = this.getUnix();          //当前时间戳
        var today = this.getTodayUnix();   //今天0点时间戳
        var year = this.getYearUnix();     //今年0点时间戳
        var timer = (now - timestamp) / 1000;   // 转换为秒级时间戳
        var tip = '';

        if (timer <= 0) {
            tip = '刚刚';
        } else if (Math.floor(timer/60) <= 0) {
            tip = '刚刚';
        } else if (timer < 3600) {
            tip = Math.floor(timer/60) + '分钟前';
        } else if (timer >= 3600 && (timestamp - today >= 0) ) {
            tip = Math.floor(timer/3600) + '小时前';
        } else if (timer/86400 <= 31) {
            tip = Math.ceil(timer/86400) + '天前';
        } else {
            tip = this.getLastDate(timestamp);
```

```
            }
            return tip;
        }
    };

    export default {
        bind: function (el, binding) {
            el.innerHTML = Time.getFormatTime(binding.value * 1000);
            el.__timeout__ = setInterval(function () {
                el.innerHTML = Time.getFormatTime(binding.value * 1000);
            }, 60000);
        },
        unbind: function (el) {
            clearInterval(el.__timeout__);
            delete el.__timeout__;
        }
    }
```

关于自定义指令和 v-time 的详细介绍，可以回顾 8.2.2 小节的内容。

评论列表在获取完文章内容后再获取，代码如下：

```
// components/daily-article.vue，部分代码省略
<template>
    <div class="daily-article">
        <div class="daily-article-title">{{ data.title }}</div>
        <div class="daily-article-content" v-html="data.body"></div>

        <div class="daily-comments" v-show="comments.length">
            <span>评论（{{ comments.length }}）</span>
            <div class="daily-comment" v-for="comment in comments">
                <div class="daily-comment-avatar">
                    <img :src="comment.avatar">
                </div>
                <div class="daily-comment-content">
                    <div class="daily-comment-name">{{ comment.author }}</div>
                    <div class="daily-comment-time" v-time="comment.time"></div>
                    <div class="daily-comment-text">{{ comment.content }}</div>
                </div>
            </div>
        </div>
    </div>
</template>
```

```
<script>
    import Time from '../directives/time';
    import $ from '../libs/util';
    export default {
        directives: { Time },
        props: {
            id: {
                type: Number,
                default: 0
            }
        },
        data () {
            return {
                data: {},
                comments: []
            }
        },
        methods: {
            getArticle () {
                $.ajax.get('news/' + this.id).then(res => {
                    // ...
                    this.getComments();
                })
            },
            getComments () {
                this.comments = [];
                $.ajax.get('story/' + this.id + '/short-comments').then(res => {
                    this.comments = res.comments.map(comment => {
                        //将头像的图片地址转为代理地址
                        comment.avatar = $.imgPath + comment.avatar;
                        return comment;
                    });
                })
            }
        }
    };
</script>

// style.css
.daily-comments{
    margin: 10px 0;
}
.daily-comments span{
```

```css
    display: block;
    margin: 10px 0;
    font-size: 20px;
}
.daily-comment{
    overflow: hidden;
    margin-bottom: 20px;
    padding-bottom: 20px;
    border-bottom: 1px dashed #e3e8ee;
}
.daily-comment-avatar{
    width: 50px;
    height: 50px;
    float: left;
}
.daily-comment-avatar img{
    width: 100%;
    height: 100%;
    border-radius: 3px;
}
.daily-comment-content{
    margin-left: 65px;
}
.daily-comment-name{

}
.daily-comment-time{
    color: #9ea7b4;
    font-size: 14px;
    margin-top: 5px;
}
.daily-comment-text{
    margin-top: 10px;
}
```

以上就是日报项目的所有细节分析和代码。

13.4 总　　结

本章所有的代码已上传至 GitHub，访问下面的链接可以查看到并直接使用：

https://github.com/icarusion/vue-book

vue-book 下的 daily 目录就是本章的代码，在该目录下执行 npm install 命令会自动安装所有的依赖，然后执行 npm run dev 启动 webpack 服务，执行 node proxy.js 启动代理服务。

日报项目以单页面的形式呈现，基本覆盖了 Vue 和 webpack 的核心功能，它们包括：

- Vue 的单文件组件用法。
- Vue 的基本指令、自定义指令。
- 数据的获取、整理、可视化。
- prop、事件、子组件索引。
- ES6 模块。

日报项目是一个较独立的单页小应用，没有使用路由和大规模状态管理插件 Vuex，在工程上并不算复杂，比较适合刚入手 Vue 的练习项目。虽然看似简单，但它覆盖了业务中很多场景，对代码进行了组织和模块化，很接近真实的生产项目。项目对代码维护和扩展性也有考虑，比如对 Ajax 的封装、通用工具函数的提取、组件的解耦等，这些细节都是在实际项目中要考虑的。

练习：参考知乎日报移动 App 的 UI 设计，基于本章内容开发移动 Web 版。

第 14 章

实战:电商网站项目开发

本章将结合本书所有的知识点(包括 webpack、Vuex、vue-router 等)来开发一个具有代表性的电商网站项目。所涉及的内容涵盖了许多典型场景,如商品列表按照价格、销量排序;商品列表按照品牌、价格过滤;动态的购物车;使用优惠码等。

本章仍然会使用前面章节的基本 webpack 配置,所有的代码也上传至 GitHub,访问链接可以查看并直接使用:https://github.com/icarusion/vue-book/tree/master/shopping。

14.1 项目工程搭建

新建目录 shopping,复制第 10.2 节的 webpack 开发环境和生产环境的两个配置文件(webpack.config.js、webpack.prod.config.js)及 package.json 等核心文件,并通过 NPM 完成安装。项目的主要配置在于 main.js 文件。

本章会使用到 Vue.js 的路由插件 vue-router 和状态管理插件 Vuex,首先在 main.js 中导入并做初始化配置:

```
// main.js
import Vue from 'vue';
import VueRouter from 'vue-router';
import Routers from './router.js';
import Vuex from 'vuex';
import App from './app.vue';
import './style.css';

Vue.use(VueRouter);
Vue.use(Vuex);
```

```js
// 路由配置
const RouterConfig = {
    // 使用 HTML 5 的 History 路由模式
    mode: 'history',
    routes: Routers
};
const router = new VueRouter(RouterConfig);

router.beforeEach((to, from, next) => {
    window.document.title = to.meta.title;
    next();
});

router.afterEach((to, from, next) => {
    window.scrollTo(0, 0);
});

const store = new Vuex.Store({
    state: {

    },
    getters: {

    },
    mutations: {

    },
    actions: {

    }
});

new Vue({
    el: '#app',
    router: router,
    store: store,
    render: h => {
        return h(App)
    }
});
```

其中，路由的页面配置放在了 router.js 文件内单独维护；Vuex 默认设置了 state、getters、mutations、actions、之后随项目需求持续添加。

项目中全局使用的一些 CSS 样式写在了 style.css 文件内，在 main.js 中直接导入，webpack 打包时，会将此 CSS 文件与.vue 单文件中的 CSS （包括 scoped）一同提取，输出到 main.css 文件。

在 shopping 目录下新建 views 目录，用于放置每个路由页面的.vue 文件；新建 components 目录用来存放公共组件；新建 images 目录用来存放项目中用到的图片。配置完这些后，通过 NPM 运行 npm run dev 命令，启动 webpack 服务，这样就完成了基础工程的搭建。

14.2 商品列表页

14.2.1 需求分析与模块拆分

商品列表页面用于展示相关的所有商品，一般具有筛选和排序两种过滤方法。比如可以按照品牌筛选（如 Apple、Beats、Bose）或颜色筛选（如白色、金色），筛选条件可以叠加（比如白色的 Beats 品牌）。可以按照价格、销量等在筛选的基础上再进行排序，最终过滤出符合要求的商品。最终完成的效果如图 14-1 所示。

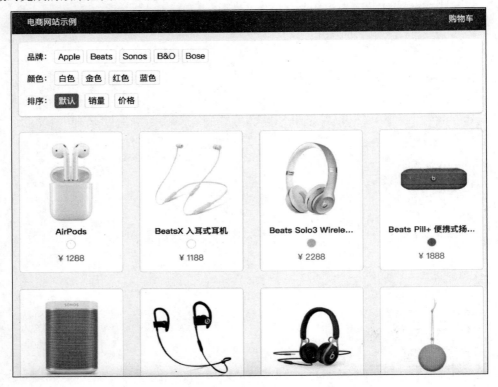

图 14-1　商品列表页效果图

排序为单选，初始按"默认"进行排序，其中价格可分为升序（价格从低到高）和降序（价格从高到低）两种排序，销量则只有降序。

第 14 章　实战：电商网站项目开发

品牌和颜色都是单选，单次点击选中，再次点击取消选中。

初次打开商品列表页会请求一次远程数据（示例用 setTimeout 模拟异步，真实场景应该通过 Ajax 获取），获取到全量的商品数据，然后筛选和排序都是在本地完成（真实场景也有在服务端进行筛选和排序的做法，因为商品很有可能会分页，前端一次性拿到所有数据不现实）。

商品列表页主要有两个模块，一个是路由组件（views/list.vue），负责数据的请求、过滤相关的逻辑；另一个是商品简介组件（components/product.vue，即每个商品卡片），鼠标经过时，显示出加入购物车的按钮，如图 14-2 所示。

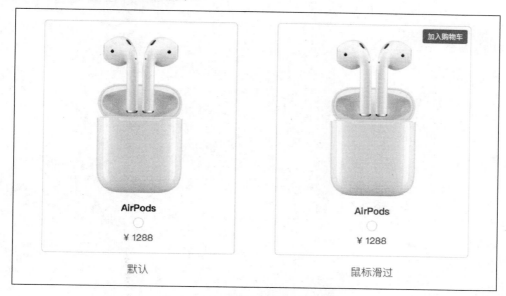

图 14-2　商品简介组件示意图

两个模块的样式都直接写在各自的 .vue 文件的 <style scoped> 部分。

14.2.2　商品简介组件

上一节中，图 14-2 已经展示了商品简介组件的效果，本节将完成该组件的开发。在 components 目录下新建 product.vue 文件。每个商品的选项比较多，比如标题、价格、颜色等，为方便父子组件之间传递，直接在 product.vue 中设置一个 property：info 来接收一个对象格式的数据，这样扩展性较高，父级也可直接将获取到的数据传递过来，省去了拆分的工作。

info 数据结构如下：

```
// info
{
    id: 1,
    name: 'AirPods',
    brand: 'Apple',
    image: 'http://ordfm6aah.bkt.clouddn.com/shop/1.jpeg',
    sales: 10000,
    cost: 1288,
```

```
    color: '白色'
}
```

其中，id 是商品的 id，点击卡片会进入该商品的详情页面，后面章节会陆续介绍。name 是商品名称，brand 为品牌，image 是图片，sales 是销量，cost 为单价，color 为颜色。颜色较特殊，因为直接返回的中文无法对应到具体的色值，所以在 product.vue 的 data 选项中定义一个 map，用于映射颜色和色值。相关代码如下：

```
// product.vue，部分代码省略
<script>
    export default {
        props: {
            info: Object
        },
        data () {
            return {
                colors: {
                    '白色': '#ffffff',
                    '金色': '#dac272',
                    '蓝色': '#233472',
                    '红色': '#f2352e'
                }
            }
        }
    };
</script>
```

鼠标悬停在卡片上时会显示"加入购物车"按钮，本节先定义好内容，具体实现将在 14.4 节完成。

product 组件的模板代码如下：

```
// product.vue，部分代码省略
<template>
    <div class="product">
        <router-link
            :to="'/product/' + info.id"
            class="product-main">
            <img :src="info.image">
            <h4>{{ info.name }}</h4>
            <div
                class="product-color"
                :style="{ background: colors[info.color]}"></div>
            <div class="product-cost">¥ {{ info.cost }}</div>
            <div
                class="product-add-cart"
```

```
            @click.prevent="handleCart">加入购物车</div>
        </router-link>
    </div>
</template>
<script>
    export default {
        methods: {
            handleCart () {
                this.$store.commit('addCart', this.info.id);
            }
        }
    };
</script>
```

<router-link>最终会渲染为一个<a>标签,链接到:to 定义的 url,也就是商品详情页,id 会作为参数通过 vue-router 传递。

"加入购物车"按钮对@click 事件使用了 prevent 修饰符来阻止冒泡,否则在点击按钮的同时,也会点击到<a> 标签进入详情页。

"加入购物车"按钮先设置了一个 handleCart 方法,通过 Vuex 触发 mutation 保存到购物车,参数为商品的 id,具体逻辑将在后面介绍。

product 组件的样式代码如下:

```
//product.vue,部分代码省略
<style scoped>
    .product{
        width: 25%;
        float: left;
    }
    .product-main{
        display: block;
        margin: 16px;
        padding: 16px;
        border: 1px solid #dddee1;
        border-radius: 6px;
        overflow: hidden;
        background: #fff;
        text-align: center;
        position: relative;
    }
    .product-main img{
        width: 100%;
    }
    h4{
```

```css
        color: #222;
        overflow: hidden;
        text-overflow: ellipsis;
        white-space: nowrap;
    }
    .product-main:hover h4{
        color: #0070c9;
    }
    .product-color{
        display: block;
        width: 16px;
        height: 16px;
        border: 1px solid #dddee1;
        border-radius: 50%;
        margin: 6px auto;
    }
    .product-cost{
        color: #de4037;
        margin-top: 6px;
    }
    .product-add-cart{
        display: none;
        padding: 4px 8px;
        background: #2d8cf0;
        color: #fff;
        font-size: 12px;
        border-radius: 3px;
        cursor: pointer;
        position: absolute;
        top: 5px;
        right: 5px;
    }
    .product-main:hover .product-add-cart{
        display: inline-block;
    }
</style>
```

这里给<style>加了 scoped 属性，所以样式只针对 product.vue 组件生效，不会影响其他组件。在 class 的命名上，有很多种规范，这里推荐以模块为首，依次用"-"分割作用域，比如.product、.product-main、.product-main-img。如果你使用 Less 或其他 CSS 预处理做开发，写起来会很舒服，比如上面的样式可以改写为：

```
// product.vue, 改写后的样式，以 Less 为例
<style scoped lang="less">
```

```less
@prefix-cls: "product";
.@{prefix-cls}{
    width: 25%;
    float: left;
    &-main{
        display: block;
        margin: 16px;
        padding: 16px;
        border: 1px solid #dddee1;
        border-radius: 6px;
        overflow: hidden;
        background: #fff;
        text-align: center;
        position: relative;
        & img{
            width: 100%;
        }
    }
    & h4{
        color: #222;
        overflow: hidden;
        text-overflow: ellipsis;
        white-space: nowrap;
    }
    &:hover h4{
        color: #0070c9;
    }
    &-color{
        display: block;
        width: 16px;
        height: 16px;
        border: 1px solid #dddee1;
        border-radius: 50%;
        margin: 6px auto;
    }
    &-cost{
        color: #de4037;
        margin-top: 6px;
    }
    &-add-cart{
        display: none;
        padding: 4px 8px;
        background: #2d8cf0;
```

```css
            color: #fff;
            font-size: 12px;
            border-radius: 3px;
            cursor: pointer;
            position: absolute;
            top: 5px;
            right: 5px;
        }
        &:hover .@{prefix-cls}{
            display: inline-block;
        }
    }
}
</style>
```

使用 CSS 预编译的好处有很多,比如支持变量、封装相同样式为函数、循环等。但是 webpack 默认是不支持的,需要配置 less-loader。

配置 less 的方法

首先通过 NPM 安装 less 和 less-loader:

```
npm install less --save-dev
npm install less-loader --save-dev
```

然后在 webpack 中配置 less-loader,部分代码省略:

```js
module: {
    rules: [
        {
            test: /\.vue$/,
            loader: 'vue-loader',
            options: {
                loaders: {
                    less: ExtractTextPlugin.extract({
                        use: ['css-loader', 'less-loader'],
                        fallback: 'vue-style-loader'
                    }),
                    css: ExtractTextPlugin.extract({
                        use: ['css-loader', 'less-loader'],
                        fallback: 'vue-style-loader'
                    })
                }
            }
        },
        {
            test: /\.less/,
```

```js
            use: ExtractTextPlugin.extract({
                use: ['less-loader'],
                fallback: 'style-loader'
            })
        }
    ]
}
```

14.2.3 列表按照价格、销量排序

在 views 目录下新建 list.vue 文件，并在 router.js 中添加商品列表的路由配置：

```js
// router.js
const routers = [
    {
        path: '/list',
        meta: {
            title: '商品列表'
        },
        component: (resolve) => require(['./views/list.vue'], resolve)
    },
    {
        path: '*',
        redirect: '/list'
    }
];
export default routers;
```

我们先把数据搞定，再来看 list.vue。列表相关的数据都通过 Vuex 来维护，可以回顾第 11 章 11.2 节的内容。

首先需要获取商品列表的数据，获取是异步的，所以要写在 Vuex 的 actions 里。在真实场景中，数据应当是通过 Ajax 从服务端获取的，本实例用 setTimeout 来模拟异步，并用本地数据来 mock。

在根目录 shopping 下新建文件 product.js，并写入以下数据：

```js
// product.js
export default [
    {
        id: 1,
        name: 'AirPods',
        brand: 'Apple',
        image: 'http://ordfm6aah.bkt.clouddn.com/shop/1.jpeg',
        sales: 10000,
        cost: 1288,
        color: '白色'
    },
```

```
{
    id: 2,
    name: 'BeatsX 入耳式耳机',
    brand: 'Beats',
    image: 'http://ordfm6aah.bkt.clouddn.com/shop/2.jpeg',
    sales: 11000,
    cost: 1188,
    color: '白色'
},
{
    id: 3,
    name: 'Beats Solo3 Wireless 头戴式式耳机',
    brand: 'Beats',
    image: 'http://ordfm6aah.bkt.clouddn.com/shop/3.jpeg',
    sales: 5000,
    cost: 2288,
    color: '金色'
},
{
    id: 4,
    name: 'Beats Pill+ 便携式扬声器',
    brand: 'Beats',
    image: 'http://ordfm6aah.bkt.clouddn.com/shop/4.jpeg',
    sales: 3000,
    cost: 1888,
    color: '红色'
},
{
    id: 5,
    name: 'Sonos PLAY:1 无线扬声器',
    brand: 'Sonos',
    image: 'http://ordfm6aah.bkt.clouddn.com/shop/5.jpeg',
    sales: 8000,
    cost: 1578,
    color: '白色'
},
{
    id: 6,
    name: 'Powerbeats3 by Dr. Dre Wireless 入耳式耳机',
    brand: 'Beats',
    image: 'http://ordfm6aah.bkt.clouddn.com/shop/6.jpeg',
    sales: 12000,
    cost: 1488,
```

```
        color: '金色'
    },
    {
        id: 7,
        name: 'Beats EP 头戴式耳机',
        brand: 'Beats',
        image: 'http://ordfm6aah.bkt.clouddn.com/shop/7.jpeg',
        sales: 25000,
        cost: 788,
        color: '蓝色'
    },
    {
        id: 8,
        name: 'B&O PLAY BeoPlay A1 便携式蓝牙扬声器',
        brand: 'B&O',
        image: 'http://ordfm6aah.bkt.clouddn.com/shop/8.jpeg',
        sales: 15000,
        cost: 1898,
        color: '金色'
    },
    {
        id: 9,
        name: 'Bose® QuietComfort® 35 无线耳机',
        brand: 'Bose',
        image: 'http://ordfm6aah.bkt.clouddn.com/shop/9.jpeg',
        sales: 14000,
        cost: 2878,
        color: '蓝色'
    },
    {
        id: 10,
        name: 'B&O PLAY Beoplay H4 无线头戴式耳机',
        brand: 'B&O',
        image: 'http://ordfm6aah.bkt.clouddn.com/shop/10.jpeg',
        sales: 9000,
        cost: 2298,
        color: '金色'
    }
]
```

在 main.js 中导入数据,并在 Vuex 中声明数据列表相关的 state、mutations、actions:

```
// main.js,部分代码省略
// 导入数据
```

```js
import product_data from './product.js';

const store = new Vuex.Store({
    state: {
        // 商品列表数据
        productList: [],
        //购物车数据
        cartList: []
    },
    mutations: {
        //添加商品列表
        setProductList (state, data) {
            state.productList = data;
        }
    },
    actions: {
        //请求商品列表
        getProductList (context) {
            // 真实环境通过 Ajax 获取,这里用异步模拟
            setTimeout(() => {
                context.commit('setProductList', product_data);
            }, 500);
        }
    }
});
```

首先通过 action 的 getProductList 方法获取数据,然后由 mutation 的 setProduction 方法将数据设置到 productList。

准备好了数据,再来看视图部分。先在根实例 app.vue 中挂载路由并设置导航条:

```html
// app.vue
<template>
    <div>
        <div class="header">
            <router-link
                to="/list"
                class="header-title">电商网站示例</router-link>
            <div class="header-menu">
                <router-link to="/cart" class="header-menu-cart">
                    购物车
                    <span v-if="cartList.length">{{ cartList.length }}</span>
                </router-link>
            </div>
        </div>
```

```
        <router-view></router-view>
    </div>
</template>
<script>
    export default {
        computed: {
            cartList () {
                return this.$store.state.cartList;
            }
        }
    }
</script>
```

数据 cartList 是购物车中添加的商品，后面会介绍。路由视图 <router-view> 挂载了所有的路由组件。

app.vue 的样式在 style.css 中全局定义：

```
// style.css
*{
    margin: 0;
    padding: 0;
}
a{
    text-decoration: none;
}
body{
    background: #f8f8f9;
}
.header{
    height: 48px;
    line-height: 48px;
    background: rgba(0,0,0,.8);
    color: #fff;
}
.header-title{
    padding: 0 32px;
    float: left;
    color: #fff;
}
.header-menu{
    float: right;
    margin-right: 32px;
}
```

```css
.header-menu-cart{
    color: #fff;
}
.header-menu-cart span{
    display: inline-block;
    width: 16px;
    height: 16px;
    line-height: 16px;
    text-align: center;
    border-radius: 50%;
    background: #ff5500;
    color: #fff;
    font-size: 12px;
}
```

商品列表页 list.vue 在初始化时调用 Vuex 的 action 触发请求数据操作，并设置计算属性从 Vuex 中读取数据 productList。相关代码如下：

```html
// list.vue，部分代码省略
<template>
    <div v-show="list.length">
        <Product v-for="item in list" :info="item" :key="item.id"></Product>
    </div>
</template>
<script>
    // 导入商品简介组件
    import Product from '../components/product.vue';
    export default {
        components: { Product },
        computed: {
            list () {
                // 从 Vuex 获取商品列表数据
                return this.$store.state.productList;
            }
        },
        mounted () {
            // 初始化时，通过 Vuex 的 action 请求数据
            this.$store.dispatch('getProductList');
        }
    }
</script>
```

打开浏览器，此时已经可以渲染出商品列表了。

实现按照价格、销量排序，就不能直接使用数据 list，也不能直接重置 list（因为过滤不是一次性的，所以不能破坏原数据，否则无法复原），所以用计算属性来动态返回过滤后的数据。相关代码如下：

```
// list.vue，部分代码省略
<template>
    <div v-show="list.length">
        <Product
            v-for="item in filteredAndOrderedList"
            :info="item"
            :key="item.id"></Product>
        <div
            class="product-not-found"
            v-show="!filteredAndOrderedList.length">暂无相关商品</div>
    </div>
</template>
<script>
    import Product from '../components/product.vue';
    export default {
        components: { Product },
        data () {
            return {
                // 排序依据，可选值为：
                // sales（销量）
                // cost-desc（价格降序）
                // cost-asc（价格升序）
                order: ''
            }
        },
        computed: {
            list () {
                return this.$store.state.productList;
            },
            filteredAndOrderedList () {
                // 复制原始数据
                let list = [...this.list];
                // todo 按品牌过滤
                // todo 按颜色过滤
                // 排序
                if (this.order !== '') {
                    if (this.order === 'sales') {
                        list = list.sort((a, b) => b.sales - a.sales);
                    } else if (this.order === 'cost-desc') {
```

```
                    list = list.sort((a, b) => b.cost - a.cost);
                } else if (this.order === 'cost-asc') {
                    list = list.sort((a, b) => a.cost - b.cost);
                }
            }
            return list;
        }
    }
}
</script>
<style scoped>
    .product-not-found{
        text-align: center;
        padding: 32px;
    }
</style>
```

ES 6 语法提示：
展开运算符，let list = [...list] 相当于克隆了一份数据。

计算属性 filteredAndOrderedList 将 list 进一步过滤，返回筛选、排序后的数据，排序依据于 data：order，默认为空，即默认的排序为 sales、cost-desc、cost-asc 时则分别按照销量、价格降序、价格升序来排序。排序直接使用 JavaScript 数组的 sort 方法对前后两个值比较大小。

把<Product>循环的数据由 list 改为 filteredAndOrderedList 后，显示的就是过滤后的数据。剩余工作只要在视图中通过操作改变 order 即可。

在模板里加入排序按钮，并绑定相关事件：

```
// list.vue，部分代码省略
<template>
    <div v-show="list.length">
        <div class="list-control">
            <div class="list-control-order">
                <span>排序：</span>
                <span
                    class="list-control-order-item"
                    :class="{on: order === ''}"
                    @click="handleOrderDefault">默认</span>
                <span
                    class="list-control-order-item"
                    :class="{on: order === 'sales'}"
                    @click="handleOrderSales">
                    销量
                    <template v-if="order === 'sales'">↓</template>
```

```html
                </span>
                <span
                    class="list-control-order-item"
                    :class="{on: order.indexOf('cost') > -1}"
                    @click="handleOrderCost">
                    价格
                    <template v-if="order === 'cost-asc'">↑</template>
                    <template v-if="order === 'cost-desc'">↓</template>
                </span>
            </div>
        </div>
    </div>
</template>
<script>
    export default {
        data () {
            return {
                order: ''
            }
        },
        methods: {
            handleOrderDefault () {
                this.order = '';
            },
            handleOrderSales () {
                this.order = 'sales';
            },
            handleOrderCost () {
                if (this.order === 'cost-desc') {
                    this.order = 'cost-asc';
                } else {
                    this.order = 'cost-desc';
                }
            }
        }
    }
</script>
<style scoped>
    .list-control{
        background: #fff;
        border-radius: 6px;
        margin: 16px;
        padding: 16px;
```

```css
        box-shadow: 0 1px 1px rgba(0,0,0,.2);
    }
    .list-control-filter{
        margin-bottom: 16px;
    }
    .list-control-filter-item,
    .list-control-order-item {
        cursor: pointer;
        display: inline-block;
        border: 1px solid #e9eaec;
        border-radius: 4px;
        margin-right: 6px;
        padding: 2px 6px;
    }
    .list-control-filter-item.on,
    .list-control-order-item.on{
        background: #f2352e;
        border: 1px solid #f2352e;
        color: #fff;
    }
</style>
```

"默认"和"销量"只能单次点击,"价格"按钮可以点击切换为升序和降序两种状态。通过判断 order 的状态,给 3 个按钮动态绑定了 class（.on）来高亮显示当前排序的按钮。刷新页面,点击切换排序状态,商品列表已经可以动态更新了。

14.2.4　列表按照品牌、颜色筛选

首先准备数据。

品牌和颜色的数据可以作为 getters 从 Vuex 的 productList 里遍历获取,示例代码如下:

```js
// main.js,部分代码省略
//数组排重
function getFilterArray (array) {
    const res = [];
    const json = {};
    for (let i = 0; i < array.length; i++){
        const _self = array[i];
        if(!json[_self]){
            res.push(_self);
            json[_self] = 1;
        }
    }
    return res;
```

```js
    }

    const store = new Vuex.Store({
        state: {
            productList: []
        },
        getters: {
            brands: state => {
                const brands = state.productList.map(item => item.brand);
                return getFilterArray(brands);
            },
            colors: state => {
                const colors = state.productList.map(item => item.color);
                return getFilterArray(colors);
            }
        }
    });
```

使用 map 方法把 productList 里的 brand 或 color 数据过滤出来，然后用 getFilterArray 方法对数组去重。

getters 里的 brands 和 colors 依赖数据 productList，与计算属性原理类似，所以只要维护好 productList、brands 和 colors 就可以自动更新。

然后在 list.vue 中把 Vuex 里的品牌和颜色数据引入，并完成列表的过滤：

```html
// list.vue，部分代码省略
<script>
    export default {
        computed: {
            list () {
                return this.$store.state.productList;
            },
            brands () {
                return this.$store.getters.brands;
            },
            colors () {
                return this.$store.getters.colors;
            },
            filteredAndOrderedList () {
                let list = [...this.list];
                //按品牌过滤
                if (this.filterBrand !== '') {
                    list = list.filter(item => item.brand === this.filterBrand);
                }
                //按颜色过滤
```

```
            if (this.filterColor !== '') {
                list = list.filter(item => item.color === this.filterColor);
            }
            // 排序...
            return list;
        }
    },
    data () {
        return {
            filterBrand: '',
            filterColor: ''
        }
    }
}
</script>
```

品牌和颜色都是单选，但是可以协同过滤。最后只需要根据操作设置正确的 filterBrand 和 filterColor，商品列表就可以自动完成对品牌、颜色的筛选以及价格、销量的排序。相关代码如下：

```
// list.vue，部分代码如下
<template>
    <div v-show="list.length">
        <div class="list-control">
            <div class="list-control-filter">
                <span>品牌：</span>
                <span
                    class="list-control-filter-item"
                    :class="{on: item === filterBrand}"
                    v-for="item in brands"
                    @click="handleFilterBrand(item)">{{ item }}</span>
            </div>
            <div class="list-control-filter">
                <span>颜色：</span>
                <span
                    class="list-control-filter-item"
                    :class="{on: item === filterColor}"
                    v-for="item in colors"
                    @click="handleFilterColor(item)">{{ item }}</span>
            </div>
        </div>
    </div>
</template>
<script>
    export default {
```

```
            methods: {
                // 筛选品牌
                handleFilterBrand (brand) {
                    //单次点击选中，再次点击取消选中
                    if (this.filterBrand === brand) {
                        this.filterBrand = '';
                    } else {
                        this.filterBrand = brand;
                    }
                },
                // 筛选颜色
                handleFilterColor (color) {
                    if (this.filterColor === color) {
                        this.filterColor = '';
                    } else {
                        this.filterColor = color;
                    }
                }
            }
        }
</script>
```

思考：Vuex 的 getters 和组件内的 computed 很相似，其实把示例中的 brands 和 colors 写在 list.vue 的 computed 中也是可以的，那么到底什么时候把数据存在 Vuex 恰当，而什么时候在组件内维护好呢？如果在业务中比较纠结，可以结合以下几点综合考虑：

- 如果数据还有其他组件复用，建议放在 Vuex。
- 如果需要跨多级组件传递数据，建议放在 Vuex。
- 需要持久化的数据（如登录后用户的信息），建议放在 Vuex。
- 跟当前业务组件强相关的数据（如示例中的 filterBrand、filterColor，它们只在当前组件有用），可以放在组件内。

14.3 商品详情页

在 views 目录下新建 product.vue 文件，并在 router.js 中添加商品详情的路由配置：

```
// router.js，部分代码省略
const routers = [
    {
        path: '/product/:id',
        meta: {
            title: '商品详情'
        },
```

```
        component: (resolve) => require(['./views/product.vue'], resolve)
    }
];
```

商品详情的路由接收一个参数 id，即商品的 id。常见的业务场景中，会以 id 作为接口的索引，查询出所有相关的数据。为了使业务更好地解耦，从商品列表页跳转至详情页时，只传递一个商品的 id，不需要其他任何数据（虽然像商品名称、价格等数据在详情页已拿到，但不传递，重新获取）。

通过 $route 可以获取当前路由的参数，并在页面初始化时请求该商品的数据，示例使用 setTimeout 来模拟异步，真实场景下应该通过 Ajax 来请求数据。我们从数据源（product.js）里通过数组的 find() 方法拿到指定 id 的数据，完成数据 mock。

```
// views/product.vue，部分代码省略
<script>
    // 导入本地数据做匹配用，真实场景并不需要
    import product_data from '../product.js';
    export default {
        data () {
            return {
                // 获取路由中的参数
                id: parseInt(this.$route.params.id),
                product: null
            }
        },
        methods: {
            getProduct () {
                // 真实环境通过 Ajax 获取，这里用异步模拟
                setTimeout(() => {
                    this.product = product_data
                        .find(item => item.id === this.id);
                }, 500);
            }
        },
        mounted () {
            // 初始化时，请求数据
            this.getProduct();
        }
    }
</script>
```

ES 6 语法提示：
数组的 find() 方法返回数组中满足提供的测试函数的第一个元素的值，示例是将其与箭头函数连用。

然后将数据写入模板即可。需要注意的是，电商网站的详情页一般为自定义的文本和图片，商家通过富文本编辑器以可视化的形式编辑好商品内容，接口返回的是 html 片段，可以直接用 v-html 指令渲染 html 内容，但在服务端要对提交的 html 做处理，避免发生 XSS 攻击。本实例将 10 张产品的图片依次展示作为商品的内容。其代码如下：

```
// views/product.vue，部分代码省略
<template>
    <div v-if="product">
        <div class="product">
            <div class="product-image">
                <img :src="product.image">
            </div>
            <div class="product-info">
                <h1 class="product-name">{{ product.name }}</h1>
                <div class="product-cost">¥ {{ product.cost }}</div>
                <div class="product-add-cart"
                    @click="handleAddToCart">加入购物车</div>
            </div>
        </div>
        <div class="product-desc">
            <h2>产品介绍</h2>
            <img v-for="n in 10"
                :src="'http://ordfm6aah.bkt.clouddn.com/shop/' + n + '.jpeg'">
        </div>
    </div>
</template>
<script>
    export default {
        methods: {
            // 加入购物车
            handleAddToCart () {
                this.$store.commit('addCart', this.id);
            }
        }
    }
</script>
<style scoped>
    .product{
        margin: 32px;
        padding: 32px;
        background: #fff;
        border: 1px solid #dddee1;
```

```css
    border-radius: 10px;
    overflow: hidden;
}
.product-image{
    width: 50%;
    height: 550px;
    float: left;
    text-align: center;
}
.product-image img{
    height: 100%;
}
.product-info{
    width: 50%;
    padding: 150px 0 250px;
    height: 150px;
    float: left;
    text-align: center;
}
.product-cost{
    color: #f2352e;
    margin: 8px 0;
}
.product-add-cart{
    display: inline-block;
    padding: 8px 64px;
    margin: 8px 0;
    background: #2d8cf0;
    color: #fff;
    border-radius: 4px;
    cursor: pointer;
}
.product-desc{
    background: #fff;
    margin: 32px;
    padding: 32px;
    border: 1px solid #dddee1;
    border-radius: 10px;
    text-align: center;
}
.product-desc img{
    display: block;
    width: 50%;
```

```
        margin: 32px auto;
        padding: 32px;
        border-bottom: 1px solid #dddee1;
    }
</style>
```

加入购物车的功能与列表页相同,将在下一节重点介绍。最后渲染的效果如图 14-3 所示。

图 14-3 商品详情页部分效果

14.4 购 物 车

最后也是购物最重要的一个环节,就是在购物车完成结算,购物车的效果如图 14-4 所示。

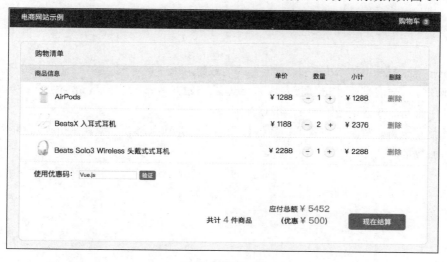

图 14-4 购物车

在购物车中,每件商品最少要选 1 件,不过可以删除,每件商品会有价格小计(单价乘以数量)。可以使用优惠码,使用后在总价的基础上减少 500 元,总价会根据购买商品的数量动态计算。右上角的购物车入口也会显示当前购物车商品的数量。

14.4.1 准备数据

之前已经提到过,将商品加入购物车是通过 Vuex 来完成的。在 main.js 中,先来定义 Vuex 中的 state 和 mutations:

```js
// main.js,部分代码省略
const store = new Vuex.Store({
    state: {
        productList: [],
        cartList: []
    },
    mutations: {
        //添加到购物车
        addCart (state, id) {
            // 先判断购物车是否已有,如果有,数量+1
            const isAdded = state.cartList.find(item => item.id === id);
            if (isAdded) {
                isAdded.count ++;
            } else {
                state.cartList.push({
                    id: id,
                    count: 1
                })
            }
        }
    }
});
```

数据 cartList 中保存购物车记录,数据格式为数组,每项是对象,包含商品 id 和购买数量两个数据(遵循解耦,其余信息通过 id 间接获取)。addCart 方法接收参数为商品 id,添加前先判断 cartList 中是否已存在商品,存在则数量加 1,不存在则写入。

有了购物车数据,剩余工作就是把数据显示出来,并动态修改数据。在 app.vue 中定义购物车入口和已添加数量:

```html
// app.vue
<template>
    <div>
        <div class="header">
            <router-link
                to="/list"
```

```html
                class="header-title">电商网站示例</router-link>
            <div class="header-menu">
                <router-link to="/cart" class="header-menu-cart">
                    购物车
                    <span v-if="cartList.length">{{ cartList.length }}</span>
                </router-link>
            </div>
        </div>
        <router-view></router-view>
    </div>
</template>
<script>
    export default {
        computed: {
            cartList () {
                return this.$store.state.cartList;
            }
        }
    }
</script>
```

在 views 目录中新建 cart.vue 文件，并添加购物车路由：

```js
// router.js，部分代码省略
const routers = [
    {
        path: '/cart',
        meta: {
            title: '购物车'
        },
        component: (resolve) => require(['./views/cart.vue'], resolve)
    }
];
```

cart.vue 中，可以先准备好以下动态数据：

- Vuex 中的购物车数据 cartList。
- product.js 中所有的商品数据（mock 用）。
- 将 product.js 中的数组转换为字典 productDictList，方便快速选取。
- 商品总数 countAll。
- 总费用（不含优惠码）costAll。

```js
// cart.vue，部分代码省略
<script>
    import product_data from '../product.js';
```

```js
export default {
    computed: {
        cartList () {
            return this.$store.state.cartList;
        },
        productDictList () {
            const dict = {};
            this.productList.forEach(item => {
                dict[item.id] = item;
            });
            return dict;
        },
        countAll () {
            let count = 0;
            this.cartList.forEach(item => {
                count += item.count;
            });
            return count;
        },
        costAll () {
            let cost = 0;
            this.cartList.forEach(item => {
                cost += this.productDictList[item.id].cost * item.count;
            });
            return cost;
        }
    },
    data () {
        return {
            productList: product_data
        }
    }
}
</script>
```

这些数据都使用了计算属性，因为彼此互相依赖。

productDictList 是对象，key 是商品 id，value 是商品信息，数据即为 product.js 中每项的内容，通过 id 可以快速便捷地获取对应商品信息。

14.4.2 显示和操作数据

在下单前，可以对每个商品的数量进行加减，或者删除商品。先将购物车数据 cartList 循环渲染，并完成表格的样式。

```vue
// cart.vue,部分代码省略
<template>
    <div class="cart">
        <div class="cart-header">
            <div class="cart-header-title">购物清单</div>
            <div class="cart-header-main">
                <div class="cart-info">商品信息</div>
                <div class="cart-price">单价</div>
                <div class="cart-count">数量</div>
                <div class="cart-cost">小计</div>
                <div class="cart-delete">删除</div>
            </div>
        </div>
        <div class="cart-content">
            <div class="cart-content-main" v-for="(item, index) in cartList">
                <div class="cart-info">
                    <img :src="productDictList[item.id].image">
                    <span>{{ productDictList[item.id].name }}</span>
                </div>
                <div class="cart-price">
                    ¥ {{ productDictList[item.id].cost }}
                </div>
                <div class="cart-count">
                    <span
                        class="cart-control-minus"
                        @click="handleCount(index, -1)">-</span>
                    {{ item.count }}
                    <span
                        class="cart-control-add"
                        @click="handleCount(index, 1)">+</span>
                </div>
                <div class="cart-cost">
                    ¥ {{ productDictList[item.id].cost * item.count }}
                </div>
                <div class="cart-delete">
                    <span
                        class="cart-control-delete"
                        @click="handleDelete(index)">删除</span>
                </div>
            </div>
            <div class="cart-empty" v-if="!cartList.length">购物车为空</div>
        </div>
    </div>
</template>
<script>
```

```
export default {
    methods: {
        handleCount (index, count) {
            if (count < 0 && this.cartList[index].count === 1) return;
            this.$store.commit('editCartCount', {
                id: this.cartList[index].id,
                count: count
            });
        },
        handleDelete (index) {
            this.$store.commit('deleteCart', this.cartList[index].id);
        }
    }
}
</script>
<style scoped>
    .cart{
        margin: 32px;
        background: #fff;
        border: 1px solid #dddee1;
        border-radius: 10px;
    }
    .cart-header-title{
        padding: 16px 32px;
        border-bottom: 1px solid #dddee1;
        border-radius: 10px 10px 0 0;
        background: #f8f8f9;
    }
    .cart-header-main{
        padding: 8px 32px;
        overflow: hidden;
        border-bottom: 1px solid #dddee1;
        background: #eee;
        overflow: hidden;
    }
    .cart-empty{
        text-align: center;
        padding: 32px;
    }
    .cart-header-main div{
        text-align: center;
        float: left;
        font-size: 14px;
    }
    div.cart-info{
```

```css
        width: 60%;
        text-align: left;
}
.cart-price{
    width: 10%;
}
.cart-count{
    width: 10%;
}
.cart-cost{
    width: 10%;
}
.cart-delete {
    width: 10%;
}
.cart-content-main{
    padding: 0 32px;
    height: 60px;
    line-height: 60px;
    text-align: center;
    border-bottom: 1px dashed #e9eaec;
    overflow: hidden;
}
.cart-content-main div{
    float: left;
}
.cart-content-main img{
    width: 40px;
    height: 40px;
    position: relative;
    top: 10px;
}
.cart-control-minus,
.cart-control-add{
    display: inline-block;
    margin: 0 4px;
    width: 24px;
    height: 24px;
    line-height: 22px;
    text-align: center;
    background: #f8f8f9;
    border-radius: 50%;
    box-shadow: 0 1px 1px rgba(0,0,0,.2);
    cursor: pointer;
}
```

```
        .cart-control-delete{
            cursor: pointer;
            color: #2d8cf0;
        }
    </style>
```

handleCount 方法用于修改购物车商品数量，最小为 1；handleDelete 方法用于删除商品。两者都根据接收的参数 index（循环 cartList 中的索引），并从数据 cartList 中获取具体商品信息。只传入 index，而不是具体数据（比如 id）的好处是更灵活、便于扩展，如果需求有所改变，就需要修改 handleCount 里的逻辑，不需要维护模板部分。

这两个方法都交给了 Vuex 中的 mutations 来操作数据：

```
// main.js，部分代码省略
const store = new Vuex.Store({
    state: {
        cartList: []
    },
    mutations: {
        //修改商品数量
        editCartCount (state, payload) {
            const product = state.cartList.find(item => item.id === payload.id);
            product.count += payload.count;
        },
        // 删除商品
        deleteCart (state, id) {
            const index = state.cartList.findIndex(item => item.id === id);
            state.cartList.splice(index, 1);
        }
    }
});
```

ES 6 语法提示：
数组的 findIndex() 方法返回数组中满足提供的测试函数的第一个元素的索引，示例是将其与箭头函数连用。

思考： 修改购物车数量时，判断是否只有 1 件的逻辑是写在 handleCount 方法内的，而不是在 Vuex 的 editCarCount 中。但事实上，写在 Vuex 里也是可以的，但不建议这样写，因为 Vuex 主要以操作数据为主，不应该关心具体的业务逻辑，业务逻辑应该在业务组件中维护。

14.4.3 使用优惠码

使用优惠码可以在总价的基础上减少指定的费用。验证优惠码的过程在真实场景也是通过 Ajax 完成的，这里仍然在本地模拟。

优惠码功能使用到两个数据：promotionCode 和 promotion，前者用于双向绑定输入框数据，后者是优惠金额。其代码如下：

```vue
// cart.vue，部分代码省略
<template>
    <div class="cart">
        <div class="cart-promotion" v-show="cartList.length">
            <span>使用优惠码：</span>
            <input type="text" v-model="promotionCode">
            <span
                class="cart-control-promotion"
                @click="handleCheckCode">验证</span>
        </div>
        <div class="cart-footer" v-show="cartList.length">
            <div class="cart-footer-desc">
                共计 <span>{{ countAll }}</span>件商品
            </div>
            <div class="cart-footer-desc">
                应付总额 <span>¥ {{ costAll - promotion }}</span>
                <br>
                <template v-if="promotion">
                    （优惠 <span>¥ {{ promotion }}</span>）
                </template>
            </div>
            <div class="cart-footer-desc">
                <div
                    class="cart-control-order"
                    @click="handleOrder">现在结算</div>
            </div>
        </div>
    </div>
</template>
<script>
    export default {
        data () {
            return {
                promotionCode: '',
                promotion: 0
            }
        },
        methods: {
            // 验证优惠码，我们用 Vue.js 代表正确的优惠码
            handleCheckCode () {
```

```
            if (this.promotionCode === '') {
                window.alert('请输入优惠码');
                return;
            }
            if (this.promotionCode !== 'Vue.js') {
                window.alert('优惠码验证失败');
            } else {
                this.promotion = 500;
            }
        },
        // 通知Vuex，完成下单
        handleOrder () {
            this.$store.dispatch('buy').then(() => {
                window.alert('购买成功');
            })
        }
    }
}
</script>
<style scoped>
    .cart-promotion{
        padding: 16px 32px;
    }
    .cart-control-promotion,
    .cart-control-order{
        display: inline-block;
        padding: 8px 32px;
        border-radius: 6px;
        background: #2d8cf0;
        color: #fff;
        cursor: pointer;
    }
    .cart-control-promotion{
        padding: 2px 6px;
        font-size: 12px;
        border-radius: 3px;
    }
    .cart-footer{
        padding: 32px;
        text-align: right;
    }
    .cart-footer-desc{
        display: inline-block;
```

```
        padding: 0 16px;
    }
    .cart-footer-desc span{
        color: #f2352e;
        font-size: 20px;
    }
</style>
```

应付总额是实际的商品总价减去优惠的价格,因为优惠价 promotion 默认是 0,可以不再判断是否使用了优惠码。

下单的操作通过 Vuex 的 action 完成,下单成功后,清空购物车数据。因为下单要通知服务端,所以需要在 action 内操作。

```
// main.js,部分代码省略
const store = new Vuex.Store({
    state: {
        cartList: []
    },
    mutations: {
        // 清空购物车
        emptyCart (state) {
            state.cartList = [];
        }
    },
    actions: {
        //购买
        buy (context) {
            // 真实环境应通过 Ajax 提交购买请求后再清空购物列表
            return new Promise(resolve=> {
                setTimeout(() => {
                    context.commit('emptyCart');
                    resolve();
                }, 500)
            });
        }
    }
});
```

在 action 中,使用 setTimeout 模拟异步,并通过返回一个 Promise 对象来通知 cart.vue 的 handleOrder 购物完成。关于在 action 中 Promise 的用法,可以回顾 11.2.3 小节 Vuex 的高级用法内容。

至此,本实战项目的所有功能已介绍完毕。

14.5 总　　结

本章所有的代码已上传至 GitHub，访问下面的链接可以查看到并直接使用：

https://github.com/icarusion/vue-book

vue-book 下的 shopping 目录就是本章的代码，在该目录下执行 npm install 命令会自动安装所有的依赖，然后执行 npm run dev 启动 webpack 服务。

在大中型项目中，尤其是多人协同开发时，最重要的是模块解耦。对于公共组件，要定义好 API（props、events、slots），公用数据要在 Vuex 或 bus 中统一维护。在业务中，要尽可能避免直接操作父链和子链来修改组件的状态，对于跨级通信最好通过 Vuex 或 bus 完成。

在协同开发时，可以将路由组件的内容拆分为多个组件，由不同的人维护，这样可以避免冲突，使模块更清晰，寻找 bug 也更有针对性。公共配置还可以使用混合（mixins）。

当项目中页面较多，在使用 Vuex 时可以将 store 分发到不同的文件或文件夹内，并参照 11.2.3 小节 Vuex 的高级用法，使用 modules 把 store 分割到不同的模板，这样对于复杂的应用更具维护性。

练习 1：将品牌和颜色的筛选扩展为支持多选，比如支持同时选择白色和红色。

练习 2：购物车数据支持持久化（本地保存，重新打开页面仍然有记录）。

第 15 章

相关开源项目介绍

本章将介绍一些实际开发中经常使用的与 Vue.js 相关的开源项目，它们包括服务端渲染框架 Nuxt.js、HTTP 库 axios 以及多语言插件 vue-i18n。使用好的开源项目可以让你的团队事半功倍。

15.1 服务端渲染与 Nuxt.js

15.1.1 是否需要服务端渲染

Vue.js 2 是支持服务端渲染的，不过在使用前有必要先了解你的业务场景和服务端渲染的特点，然后权衡是否真的需要服务端渲染。

服务端渲染（SSR）并不是什么新鲜技术，从互联网开始至今，大部分网站的内容仍然是由服务端渲染的，然后返回到客户端。如何查看一个网站是否是 SSR 呢？很简单，比如打开一个含有文章内容的网站，查看它的源代码，看这些文字是不是都在源代码里面，如果是，那它就是 SSR；或者通过 Chrome 调试工具，在 network 面板查看有没有相关的异步请求来调取内容。

很多网站之所以使用 SSR，主要目的是做搜索引擎优化（SEO）。由于所处的国家和利益不同，谷歌很早就支持对使用 Ajax 技术异步渲染内容的网站进行爬取，它们洞见了这种技术将会被广泛利用，不过谷歌在服务端的开销也要增加很多，因为这依赖于一个模拟的浏览器环境。百度至今仍不支持爬取动态渲染内容为主的网站，可能是国内目前需营销的网站大多还是静态内容站吧。因此，是否需要 SSR，最主要的因素就是是否需要 SEO，换句话说，你的产品是面向大众用户的，还是面向企业的。如果是面向企业，那可能只有首页、信息页和一些营销页面需要 SEO，与主产品分离。

使用 SSR 的第二原因是，客户端的网络可能是不稳定的，有的地方很快，有的地方会很慢。这种情况下，通过 SSR 减少请求量和客户端渲染可以相对快速地看到内容。

SSR 听起来很不错，使用它也是有前提的，那就是你的团队需要懂 Node.js 的小伙伴。Vue.js 的后端渲染不同于 PHP 的模板或 JSP 等网站，你的产品可能还是由 PHP、Java 等后端来提供数据接口，Node.js 在这里只负责渲染，也就是中间层（大前端）。如果你的团队具备了这些技术能力，产品也有 SSR 的场景，那就可以尽情地使用 Vue.js 的 SSR。

15.1.2　Nuxt.js

项目地址：https://github.com/nuxt/nuxt.js。

Nuxt.js 是一个基于 Vue.js 的通用应用框架，为 Node.js 做 Vue 的服务端渲染提供了各种配置。使用 Nuxt.js，你可以轻松、快速地搭建一套 SSR 框架，省去了大量配置的工作。

为了快速体验 Nuxt.js，可以下载安装 starter 模板（https://github.com/nuxt/starter/archive/source.zip）。下载后，通过 npm install 安装依赖，再通过 npm run dev 启动项目，在浏览器访问 127.0.0.1:3000 即可预览，如图 15-1 所示。

图 15-1　Nuxt.js 项目

与普通 Vue.js 项目不同的是，Nuxt.js 构建的代码，UI 是在服务端渲染的，而非在客户端。通过 webpack 创建的 SPA 项目，查看其源代码，<body>内一般只有一个<div>元素作为根实例挂载节点，其他都由 JavaScript 来渲染。而查看 Nuxt.js 构建后的源代码，所有模板内容直接渲染在其中。

使用 Nuxt.js 基本与写.vue 单文件一致，可到项目下的 pages 内查看首页文件 index.vue。更多 Nuxt.js 详细的用法和 Vue.js 的 SSR 可以阅读官方文档。

Vue.js 服务端渲染（SSR）文档：https://ssr.vuejs.org/。

Nuxt.js 文档：https://nuxtjs.org/。

15.2　HTTP 库 axios

项目地址：https://github.com/mzabriskie/axios。

axios 是一个基于 Promise，同时支持浏览器端和 Node.js 的 HTTP 库，常用于 Ajax 请求。

Vue.js 不像 jQuery 或 AngularJS，本身并没有携带 Ajax 方法，因此需要借助插件或第三方 HTTP 库，而 axios 就是一个很不错的选择。

可以通过 NPM 或 CDN 的形式来使用 axios，以 NPM 为例，先进行安装：

```
npm install axios --save
```

axios 提供了多种请求方式，比如直接发起 GET 或 POST 请求：

```
axios.get('/user?ID=12345')
    .then(function (response) {
        console.log(response);
    })
    .catch(function (error) {
        console.log(error);
    });


axios.post('/user', {
        firstName: 'Fred',
        lastName: 'Flintstone'
    })
    .then(function (response) {
        console.log(response);
    })
    .catch(function (error) {
        console.log(error);
    });
```

基于 Promise，可以执行多个并发请求：

```
function getUserAccount() {
    return axios.get('/user/12345');
}

function getUserPermissions() {
    return axios.get('/user/12345/permissions');
}
```

```js
axios.all([getUserAccount(), getUserPermissions()])
    .then(axios.spread(function (acct, perms) {
      //请求都完成时...
    }));
```

也可以通过写入配置的形式发起请求:

```js
axios({
    method: 'post',
    url: '/user/12345',
    data: {
        firstName: 'Fred',
        lastName: 'Flintstone'
    }
})
    .then(function (res) {
        console.log(res);
    });
```

在业务中，经常将其封装为实例的形式调用，便于做通用配置，例如:

```js
// util.js
const instance = axios.create({
    baseURL: 'https://some-domain.com/api',
    timeout: 1000,
    headers: {'Content-Type': 'application/x-www-form-urlencoded;'}
});

export default instance;


// index.vue
<template>
    <div></div>
</template>
<script>
    import Ajax from './util.js';
    export default {
        mounted () {
            Ajax({
                method: 'post',
                url: '/user',
                data: {
                    firstName: 'Fred',
                    lastName: 'Flintstone'
                }
```

```
        }).then(res => {
            console.log(res);
        });
    }
}
</script>
```

更多关于 axios 的配置,可到 GitHub 项目主页阅读其文档。

15.3　多语言插件 vue-i18n

项目地址：https://github.com/kazupon/vue-i18n。

vue-i18n 是一个 Vue.js 插件,提供了多语言解决方案。如果你的项目有多国语言的需求,可以使用它很快速地实现。

通过 NPM 来安装：

```
npm install vue-i18n --save
```

然后在 webpack 入口文件中使用插件：

```
// main.js
import Vue from 'vue';
import VueI18n from 'vue-i18n';

Vue.use(VueI18n);
```

使用 vue-i18n 插件需要在入口文件中进行多语言包的配置,其实是一个对象,每种语言对应于一个 key：

```
// main.js 接上面代码
// ...
const messages = {
    en: {
        message: {
            hello: 'hello world'
        }
    },
    cn: {
        message: {
            hello: '你好,世界'
        }
    }
}
```

```
const i18n = new VueI18n({
    locale: 'en',           // 设置当前语言
    messages,               // 设置语言包
})

new Vue({
    el: '#app',
    router: router,
    i18n:i18n,
    render: h => h(App)
});
```

配置好多语言后，在业务组件中就可以直接使用了，例如：

```
// index.vue
<template>
    <div class="index">
        <p>{{ $t("message.hello") }}</p>
    </div>
</template>
```

\<p\>元素的内容会根据当前传入的语言自动替换语言包对应 message.hello 的内容。

更多关于 vue-i18n 的配置和说明可以查阅其文档，vue-i18n 分 6.x 和 5.x 两个版本，使用会稍有不同，上面介绍的是 6.x 版本。

6.x 文档：http://kazupon.github.io/vue-i18n/en/started.html。

5.x 文档：https://kazupon.github.io/vue-i18n/old/。